Chemistry

From the atom to beyond

PART 2: Reactions

By David J Bailey, Ph.D.

Copyright by Neosho Research Group, Inc.

Neosho Research Group, Inc.
Emporia, KS 66801

All rights reserved. No part of this book may be reproduced in any form or by any means without permission in writing from the publisher.

Data tables are republished with permission of Taylor and Francis Group LLC Books, from CRC Handbook of Chemistry and Physics, 73rd edition, published 1993; permission conveyed through Copyright Clearance Center, Inc."

Table of Contents

Forward *4*
Acknowledgements *5*
Chapter 11: Water and Concentration *6*
Chapter 12: Gases *16*
Chapter 13: Heat Transfer *27*
Chapter 14: Thermodynamics *43*
Chapter 15: Kinetics *51*
Chapter 16: Equilibrium *58*
Chapter 17: Reactions Involving Acids and Bases *71*
Chapter 18: Oxidation and Reduction *87*
Chapter 19: Electrochemistry and Batteries *100*

Forward

This is the second part of two in a set of books that is designed to be a supplement to secondary chemistry curriculum. I have written this book for the high school student in mind. It is frustrating to find that text books are increasing in price and about half of the pages the students rarely glance at let alone read. I also developed this set of texts so the student can write in the text, highlight the text and annotate the text. Current textbooks are so expensive that many teachers are hesitant to recommend this.

The pictures are nice but many times extraneous. Besides the pictures, which everyone can find on the web, I have not included many problems or questions for students to answer. Also, other educational tools such as possible careers and chapter summaries are removed from this text. The internet has become a wealth of information and students can easily access this information, probably faster than I can.

There are two parts because I found that to go in depth in the topic it takes time and students have a wide range of learning pace. The high school environment leaves little time for in depth study in and after school. This is part 2, which should roughly correspond to the second year, or semester, of a chemistry class. This can also help those students involved either in AP Chemistry or IB Chemistry.

I have remove most references to old atomic theories. I understand the importance in recognizing scientists who have made progress in the field of chemistry, but often the focus is on one group of chemists and physicists. I leave the history lesson to other texts. I hope this helps the reader understand chemistry and the complexity of our world. Chemistry is a very interesting and often perplexing field to study.

Enjoy the book.

Acknowledgements

Again, I would like to thank my family for being patient and supportive during this endeavor. I would like to thank my editor, Becky DeJesus. She has helped me keep my prepositions straight and the text readable. Also, a shout out to my students who have been instrumental in telling me what has or has not worked when teaching this material. Lastly, thank you for purchasing this book.

Sincerely,
Dave Bailey, Ph.D.

Chapter 11: Water and Concentration

Why is water so important?
Water is an important compound for many reasons. For one, water is a requirement for life on our planet. Water has unique properties that make this compound special. Many of the unique properties are a result of water's ability to form hydrogen bonds with itself. This property of water helps water become a liquid at room temperature. The water's mass is very low, yet the melting and boiling point for water is extremely high compared to other compounds with similar mass. Hydrogen bonding helps explain the difference in density between solid water (ice) and liquid water. Ice has a lower density than water because the water crystal occupies more space than liquid water. Hydrogen bonding can also explain how some compounds can be soluble in water and not in other compounds.

What is a hydrogen bond?
The electrons in a covalent bond between hydrogen and a highly electronegative element, such as fluorine, oxygen, sulfur, and nitrogen, is shared unequally between the two atoms. This forms a polar bond where a small negative charge is on the oxygen and the hydrogen has a small positive charge. A hydrogen bond forms

when the slightly positive hydrogen that is bound to an oxygen, nitrogen, sulfur, or fluorine interacts with a lone pair from an oxygen, nitrogen, sulfur, or fluorine. This is a very strong interaction. The mechanics of this interaction is discussed in detail in chapter 13. At this time, the definition above will help understand how water interacts with other materials.

Why is water used as a solvent?
Water is a good solvent because of its very high polarity. The oxygen atom has a very high electronegativity (the second highest among all the elements), and since hydrogen has a single electron, the electron density is shifted towards the oxygen atom causing the hydrogen end to be partially positive and the oxygen to be slightly negative. This polarity helps to dissolve molecules and dissociate inorganic salts.

When an inorganic or ionic salt is added to water, the salt dissociates into its respective components, both positive cations and negative anions. Dissociation is a physical property, because once water is removed from the solution the original compound remains. When this occurs, the solution can transfer an electrical charge. Pure, neutral water is an electrical insulator. However, in most water systems water is a conductor due the presence of ions in the water. Hence, DO NOT place an electrical appliance in water, it will be a shocking experience! An ionic salt dissolved in a solvent is now defined as an electrolyte. A strong electrolyte is defined as completely dissociating in the solvent, and a weak electrolyte is a compound that dissolves in, and partially dissociates, in the solvent.

Can a solid contain water?
Water can also be found in the crystals of inorganic compounds or ionic salts. When the ionic solid contains water in the crystal, the solid is defined as being hydrated. In many cases, heat can be applied to the solid to drive out the water. Conversely, the desiccated or dried compound can absorb water to form the hydrate. Many drying agents work in this way.

What is a solution?
Solution is defined as the homogeneous mixture of two or more compounds; the solvent is the substance that contains the highest amount of a compound and the solute(s) is the compound or compounds that is dissolved in the solvent. There are many different types of solutions. If a gas is a solvent, we can have gasses dissolved in a gas. Air is a good example of this. Then liquid can be dissolved in a gas, think about a mist or clouds. Solids can also be dissolved in a gas: an aerosol or dust in the air are good examples of a solid dissolved in a gas. Now, with gas as a solvent, many scientists classify them as a heterogeneous mixture, not a solution which is a homogeneous solution. This depends on how small the particles are.

Liquid solvents are easier to classify as a solution. Gases are often dissolved in a liquid, think of carbonated water or any stream that contains life. Water must contain oxygen. Then liquids can dissolve other liquids, a bottle of hydrogen peroxide obtained from a store is an example. Then solids can be dissolved into the liquid. Simple syrup, which is sugar dissolved in water, is an example. Gases and liquids can be dissolved into a solid, but a solid dissolved in a solid mixture is harder to classify, because many times the mixture is actually an alloy, indicating a chemical bond has taken place. For this chapter, the solvent will always be water unless otherwise specified. Most aqueous solutions that are used in chemistry class are defined as dilute solution, where the solute is a very minor component of the solution.

What are miscible and immiscible mixtures?

When two liquids are mixed and remain in one phase, the two liquids are defined as being miscible with each other. The solvent is the compound which has the most moles in the mixture. An example of a miscible solution is ethanol in water or isopropyl alcohol in water. Two liquids are immiscible with each other when they form two phases when added together. Salad dressing that is comprised of olive oil and vinegar is an example of an immiscible mixture.

However, when two compounds are immiscible with each other it means that the bulk amounts are not mixing. Often, there is a slight solubility of the compounds in each other. For example, olive oil dissolves slightly in water, which means that a very small amount is dissolved and can be detected.

How to make a solution?

To make a solution, the solute is measured into a flask or beaker. There may be a small amount of solvent in there to begin with, depending on the solute. ALL ACIDS must be added to water! Then add enough water to make a total amount of solution. Most glassware to make solution are designed to accurately contain a specific volume – common values are 100 mL, 500 mL, and 1 L.

What does the term concentration mean?

Before we delve deeper into these properties, the next few sections are critical for the understanding of water properties. The concept of concentration is a difficult one for students. Concentration is one of those terms that has different meanings depending on the use of the word and the situation the term is being used. In chemistry, concentration is defined as the strength of the solution.

Are there units for concentration?

There are many units of concentration, usually defined as the amount of solute to amount of solvent. The common units used to describe concentration are percent by weight, percent by volume, molarity, and molality. Percent by weight and volume is used to describe a solution with a large amount of solute and is usually calculated by either of the following equations:

$\%_{mass}$ = mass of solute/mass of solvent

$\%_{volume}$ = volume of solute/volume of solvent

X = moles of solute/total moles

Where chi (X) is defined as the mole fraction.

Molarity (M) is the most common used concentration unit in chemistry. Since it is important to know the number of particles used in a chemical reaction, molarity is the number of molecules/atoms in a liter of solvent and is defined by the following equation:

$$M = \frac{moles\ (n)\ of\ solute}{volume\ (L)\ of\ solution}$$

Molality is a different unit of concentration. This is used under certain circumstances where the mass of the solvent is important for the property of the solution. Molality is the number of particles of solute in 1 kilogram of solvent. The equation is shown below:

$$m = \frac{moles\ (n)\ of\ solute}{mass\ (kg)\ of\ solvent}$$

Please be aware of the symbol for molarity (M) and molality (m). Many students confuse these terms and variables. Molarity is used more often than molality, so most problems a student sees will deal with molarity. If water is the solvent, it is common to assume that 1 L of water is equal to 1 kg of water. However, remember that for water the density changes drastically between the values of 0 – 4 °C, and ice is less dense than liquid water.

What is normality?

The concentration unit of normality (N) is an antiquated unit that is being phased out of many text books and use in the chemistry field. Normality is the number of protons per liter of solution or the number of electrons being transferred in a redox reaction using this solution. For example, a 1 M solution of HCl is 1 N, while a 1 M solution of H_2SO_4 is 2 N. The use of normality is going out of use due to the fact that molarity is a simpler unit to create solutions for any reaction.

What is solubility?

Solubility is defined as the maximum amount of solute in grams that can be dissolved in 100 mL of solvent. When a solute reaches its maximum solubility in a solvent, you will notice that the solid solute will be on the bottom of the flask. No more solute can be dissolve in the solvent at that temperature. One of the properties of solubility is the dependence on temperature. Most solutes will increase its solubility in a solvent when the temperature increases.

What is supersaturation?

A concept that is difficult to visualize is supersaturation. A solution is called supersaturated when the solute is dissolved in a solvent at a concentration higher than its solubility. This situation can be achieved if a solvent is heated to a higher temperature and the solute is added to its solubility at that temperature. Then the solvent is cooled, and in many cases the solute will remain dissolved in solution. However, this condition is very unstable. Any type of energy, even scratching the side of the container will cause the solute to fall out of solution. This act is called precipitation, and a solid will be formed on the bottom of the container.

What is meant by serial dilution?
Serial dilution is a process that creates a solution containing a small amount of solute from a solution containing a lot of solute. For example, if a container contains 500 mL of a 4 M solution of NaCl, you need to add 1 mL of this solution to a 100 mL volumetric flask and fill the flask to the line to create a 0.04 M solution of NaCl. There is an easy equation to use to determine the volume or concentration of solution:

$$M_1 V_1 = M_2 V_2$$

Where subscript 1 means the initial conditions and subscript 2 indicates final conditions. Realize the concentration times volume is equation to moles, so the moles of solute taken from the first flask must equal the amount of moles of solute in the second flask.

What does freezing point depression mean?

When a solute is added to a solvent it will interfere with the solvent's ability to form a solid, thus decreasing the temperature of freezing. The higher the number of particles in solution, the lower the freezing point due to increased amount of interaction between solute and solvent or a decrease in solvent – solvent interactions. The amount of decrease in the freezing point can be calculated by the following equation:

$$\Delta T_d = nmK_d$$

where K_d is the freezing point depression constant for the solvent, m is the molality of the solute in the solution, and n is a correction factor based on how many particles the solute contains.

    n = 1 for all molecules that are dissolved in the solvent

    n = 2 for 1:1 ionic salts – for example NaCl

    n = 3 for 1:2 or 2:1 ionic salts, such as $BaCl_2$ or K_2S

The variable n can be higher than 3.

What does boiling point elevation mean?

Like freezing point depression, a solute can affect the boiling point of a solvent. The solvent will interact with the solute and it will take more energy to create a vapor or to boil the solvent. Like freezing point depression, the number of particles dissolved or dissociated in the solvent will affect the boiling point of the solvent. The amount of increase in the boiling point can be calculated by the following equation:

$$\Delta T_b = nmK_b$$

where K_b is the boiling point elevation constant for the solvent, m is the molality of the solute in the solution, and n is a correction factor, based on how many particles the solute contains.

n = 1 for all molecules that are dissolved in the solvent

n = 2 for 1:1 ionic salts – for example NaCl

n = 3 for 1:2 or 2:1 ionic salts, such as $BaCl_2$ or K_2S

The variable n can be higher than 3.

Chapter 12: Gases

What is the theory governing gas?

Currently, the molecular explanation of gases' behavior is the kinetic molecular theory. In this theory, gases are assumed to be very small and unreactive atoms or molecules, which will move through space at an average speed. When an average speed is discussed, the molecules or atoms may be faster or slower than another; the direction of the molecules is immaterial for this chapter. The assumption will be made that the atoms or molecules will not react, but interact in a purely elastic collision. Gas reactions will be discussed later. Since the atoms are moving, the gas has kinetic energy. The kinetic energy of an object is related to average speed by the equation:

$$KE = 0.5mv^2$$

where m is the mass of the object, v is the average speed of the object, and KE is the variable for kinetic energy.

In physics, v is velocity and is a vector – meaning it has a direction component. In this chapter, we will use the average speed and ignore the direction component of velocity for now. If the gas interacts with another atom by collision, the interaction is elastic. This means that only energy can be transferred to other atoms.

What is pressure?

To investigate the properties of a gas, a scientist must review the physical properties of the gas that can vary. For this chapter, I will only describe how simple atomic gases will interact, thus eliminating any reaction that may occur. By definition, a gas has no set volume or shape. In a volume of space, any object or compound can exert a pressure.

Pressure is defined as a force exerting on a surface area. A person can exert a pressure on a table by pushing on it with a hand. Gas atoms can do the same thing by the gas particles hitting a surface. So, pressure can be related to the number of particles striking the surface during a period. The more gas particles striking the surface, higher the pressure.

How can number of particles affect the pressure?

If the number of moles of gas in the container can vary, it can either affect the volume or the pressure of the gas, depending which variable is held constant. One of our assumptions is that gas molecules have a lot of empty space between them. Therefore, it will be easy to add more gas to the container. The more particles in the container, the number of collisions will increase and the pressure will increase. Decrease the number of particles, decrease the pressure.

How can temperature relate to gases?

Temperature is used to measure kinetic energy. This relationship is described by the following equation for a mole of particles:

$$KE = \frac{3RT}{2}$$

Where KE is kinetic energy, R is the gas constant (8.314 J/mol*K), and T is temperature in Kelvin.

Why use Kelvin temperature scale?

The Kelvin temperature scale is used instead of Fahrenheit and Celsius because of the negative values Fahrenheit and Celsius scales can measure. If you use either scale in a gas law equation, there would be chance that you can end up with a negative value for volume or pressure, which will not make sense or mean anything. Lord Kelvin developed the Kelvin scale to prevent these types of calculation errors and hypothesized the concept of absolute zero.

What happens if volume and pressure vary?

With this example, we have an adjustable syringe attached to a pressure sensor, keeping temperature and amount of gas constant. If the plunger is half way up the syringe and we place the sensor at the open end, the pressure should read 1 atm. If we compress the gas by making the volume smaller, the pressure should increase. Looking at this on the molecular level, at 1 atm of pressure, the number of gas molecules hitting the pressure sensor is a steady number per second. When we decrease the volume, we are effectively reducing the distance the gas particles move from one wall to another, increasing the number of particles hitting the wall per second. Thus, the pressure increases. If we increase the volume by extending the plunger of the syringe, the gas particles must cover a longer distance and the pressure will decrease due to a decrease in the number of particles hitting the wall per second. The mathematical relationship, called Boyle's Law, is described below:

$$PV = k, \text{ or } P_1 V_1 = P_2 V_2$$

Where P is pressure, V is volume, k is an constant containing temperature, the number of atoms, and a correction factor; and the subscript 1 indicates the initial conditions and subscript 2 is the final conditions.

What happens when pressure and temperature vary?

At room temperature, the atoms are striking the sides of the container, thus applying a force on the surface of the container. That means each interaction will register as pressure on the container. So, pressure is related to the number of collisions on a surface. Increase the temperature of the container, the gas molecules increase their velocity and the number of collisions on the side of the container will increase. Thus, pressure and temperature are directly related. A scientist will write this as an equation, as shown below:

$$P = kT, \text{ or } \frac{P_1}{T_1} = \frac{P_2}{T_2}$$

where k is a constant containing volume, number of atoms, and a correction factor; P is pressure, V is volume, and the subscript 1 indicates the initial conditions and subscript 2 is the final conditions.

What happens when volume and temperature vary?

In this example, we will use a standard balloon containing a liter of gas and we will vary the temperature by placing the balloon in a freezer or a hot oven. Initially, the balloon pressure will be atmospheric pressure, which is 1 atmosphere or 1 Atm. When the balloon is placed in a -40 °C freezer, 233 K, the balloon will shrink. This is due to the fact that the molecules are slowing down due to the decrease in temperature. When the molecules slow down, the volume should decrease in size to keep the pressure constant. So, when the balloon is placed in a 200 °C oven, 473 K, the balloon will increase its volume to compensate for the increase velocity of the gas particles. Remember, the pressure of the balloon is constant, 1 Atm. In order to keep the force constant, the number of particles hitting the wall must be constant and to be constant, the distance must increase. The mathematical equation is described below:

$$V = kT, or \ \frac{V_1}{T_1} = \frac{V_2}{T_2}$$

where k is a constant containing pressure, number of atoms, and a correction factor; T is temperature in Kelvins, V is volume, and the subscript 1 indicates the initial conditions and subscript 2 is the final conditions.

What happens when pressure and the amount of gas vary?
For this example, a can of air is being filled. At the start of the experiment, the can contains some gas at 1 atm of pressure. When we add gas to the can, we are increasing the number of collisions in the can, thus increasing the pressure of the can. The walls of the container will observe an increase in the number of collisions, thus the pressure of the container will increase. The relationship can be written as shown below:

$$P = kn, or \ \frac{P_1}{n_1} = \frac{P_2}{n_2}$$

where k is a constant containing volume, temperature, and a correction factor; P is pressure, n is the number of atoms in the container, and the subscript 1 indicates the initial conditions and subscript 2 is the final conditions.

What happens when we vary the volume and the amount of gas?
With the above example, pressure and temperature is constant. Now, we will use the inflation of a balloon. As gas is added to the balloon, the volume is increased. As we add more particles to the

system, the volume must increase to keep the number of collisions constant on the walls. The mathematical relationship is below:

$$V = kn, \text{ or } \frac{V_1}{n_1} = \frac{V_2}{n_2}$$

where k is a constant containing pressure, temperature, and a correction factor; V is volume, n is the number of atoms in the container, the subscript 1 indicates the initial conditions and subscript 2 is the final conditions.

What will happen if the amount of gas and temperature varies?

This example is very difficult experimentally, because temperature of the gas can be measured, the temperature cannot affect the number of particles. Therefore, when we increase the number of gas particles in our system, keeping both pressure and volume constant, the temperature will decrease. Taking a closer look at the beginning of our experiment, we have one mole of gas having a certain amount of kinetic energy. When we add another mole of gas, the kinetic energy is decreased! Remember, we are keeping pressure constant. Since we are adding more particles to the container, the kinetic energy is decreased in order to keep the pressure constant. If we raise the temperature, to keep the pressure constant we need to remove the gas from the container, which can be difficult to do experimentally. Mathematically, the relationship is shown below:

$$Tn = k, \text{ or } T_1 n_1 = T_2 n_2$$

where k is a constant containing pressure, volume, and a correction factor; T is temperature, n is the number of atoms in the container, the subscript 1 indicates the initial conditions and subscript 2 is the final conditions.

Can we combine these variables into one equation?

These equations can be combined into one equation, PV=nkT. However, the constant is a general constant for all gases and vapors and the variable R is used for this gas constant and the value is 8.314 (L*Atm)/(mol*K). The actual equation is called the ideal gas law and is written below:

$$PV = nRT \text{ or } \frac{P_1 V_1}{n_1 T_1} = \frac{P_2 V_2}{n_2 T_2}$$

This ideal gas law can be used for gas compound or element that is in the gas phase. The units for the variable must be in atmosphere (atm) for pressure (P), liters (L) for volume (V), moles (mol) for number of particles (n), and Kelvin (K) for temperature (T). The unit for temperature is important because a negative pressure, volume, or number of atoms cannot exist.

Why isn't the line linear at high pressure and low volumes?

At high pressure the gas particles interact with each other. Since the kinetic energy is high, the substance cannot condense to a liquid or a solid, but it does change phase. This means that there is an upper limit that the ideal gas is applicable. There is a lower limit as well. If we lower the temperature of a gas, the volume will decrease until the gas condenses. Before that the molecule volume of the gas will affect the equation. To accurately describe these effects mathematically, chemists use the Van der Waals gas equation. This equation was developed to mathematically compensate for deviations with the ideal gas law equation when dealing with extremes of temperature and pressure. At low temperature, the scientist must account for the volume of the atoms, especially near the point where condensation of the gas can occur. Therefore, there is a correction value for volume V-nb, where b is the volume of the particle. For extreme pressure, a correction value is needed for the pressure, P + (an²/V²). This adjustment will correct the pressure for any intermolecular interaction that will occur at high pressure. The equation can be written below:

$$P = \frac{nRT}{(V-bn)} - \left(\frac{an^2}{V^2}\right)$$

Where a is the correction factor for the interactions between the particles and b is the correction factor for the size of the gas particle.

Other Properties of Gases

Can we describe the pressure of gases in a mixture?

Air is a mixture of gases; how does this mixture affect gas properties? For example, in a balloon nitrogen and oxygen are added so that the final composition is 50% nitrogen and 50% oxygen by mole. The mole fraction is a percentage based on the number of molecules of a compound divided by the total. At the molecular level, half of the particles hitting the side are nitrogen and the other half are oxygen. Using this example, the partial pressure of nitrogen is 0.5 of the total pressure. The equation below expresses this relationship in mathematical terms:

$$P_T = \sum P \quad \text{or} \quad P_T = P_1 + P_2 + P_3 \ldots$$

Where P is pressure and the subscripts indicate the gases in the mixture.

How can a gas escape from a balloon?

If you obtain a helium balloon and keep it for a few days, you should observe that the volume of the balloon will decrease and eventually shrink. This process can be described as effusion, which is the rate of escape of a gas from a closed container through a pore or a very tiny hole. The rate is dependent on the mass of the gas, where the heavier particles will effuse the slowest. The mathematical expression is described below:

$$rate \propto \frac{1}{\sqrt{m}}$$

This relationship is known as Graham's law. This relationship will not calculate the absolute rate, but if the rate of effusion is known for one experiment, then the rate of effusion for another gas, using the same equipment, can be calculated using the equation below:

$$\frac{rate_1}{rate_2} = \frac{\sqrt{m_2}}{\sqrt{m_1}}$$

Diffusion is similar to effusion, it is defined as the rate of migration of a gas through another gas or medium. The classic visualization of diffusion is to add a drop of food coloring into a glass of pure water and observe the water changing from clear and colorless to the color of the dye. This also happens with compounds in the gas phase. If someone opens a bad egg in a corner of the room, it will take time for the odor to reach someone on the opposite side of the room. That is diffusion. Like effusion, the rate of diffusion is dependent on the experimental conditions and the gas being used. The rate of diffusion can be compared and calculated using Graham's Law.

Chapter 13: Heat Transfer

Is the temperature of a solid the same as a gas?
In the last chapter we related the average kinetic energy of a gas to the temperature of the gas. In a liquid and solid, the temperature is also related to the kinetic energy; however, the kinetic energy is not totally translational, especially in a solid. It is found in the rotation of the molecule or atom and the vibration of the molecule's bonds. In a liquid you do have some movement of the particles, but they are interacting with other particles that slow them down. In solids all the kinetic energy the thermometer is measuring is due to the vibration and rotational energy of the particles.

What types of energy are found in chemistry?
To look at the energy of a substance, the individual atoms of elements in a solid must be studied. All atoms in a solid are moving by rotating around an axis and the electrons are moving around the atom. Therefore, each atom has rotational energy. The atoms will also vibrate within the crystal structure, and this is classified as vibrational energy. The electrons also have chemical potential energy. The actual amount can be calculated from

quantum mechanics and orbital theory. The atoms will utilize this energy to help with bonding to other atoms.

When the elements melt to form a liquid, the atoms gain translational movement, kinetic energy. The atoms will still have intermolecular forces that will interact with each other. Then, as a gas, most of the energy is kinetic, and there are minimal interactions with other molecules. Note: An interaction in chemistry is when two substances can affect the physical properties, but will not exchange atoms. A reaction is when atoms move from one substance to another to make a new substance.

Sounds complicated, but the energy is just divided into different forms to keep track of the energy. Some can be accessed during a reaction and others cannot. Rotational energy of the atom cannot be accessed during a chemical reaction, but the potential energy of the electrons can be used. Also, as we increase the temperature of the substance, the electrons also absorb some of this energy to become excited and move to a higher energy level that is virtual. A virtual orbital is an atomic or molecular orbital that has more energy than the ground state orbital but has a very short time the electron can be in that orbital. During this excited state is when some of the reactions can take place. In this chapter, gas reactions will not be discussed due to their complexity.

So, how is the total energy categorized in a molecule?

Molecules are a little more complicated due to their structures. In a solid the atoms rotate as if they were unbounded. However, since the atoms are bound to another atom, the electron cloud is not "spinning" as much, and the electrons are constrained to a region in space by the molecular orbitals. The electrons have potential chemical energy that can be accessed by other molecules during a reaction. In addition, the molecules will vibrate and the bond will stretch back and forth like a spring. This vibrational energy is usually not accessed during a reaction, but it can help a reaction to occur.

In the liquid phase molecules are moving, hence an increase in kinetic energy. Also, the molecule is rotating around an axis, increasing the total energy in the system. The kinetic energy of the molecule is not high enough to break any interaction between the individual molecules. When the molecule becomes a gas most of the energy is now kinetic, and like elements, has limited interactions with other gas phase molecules.

In any chemical system, the rotational energy of the atoms is quite small so we can safely disregard this amount. This is also true with vibrational energy and rotational energy of the molecules. In fact, the total energy of a substance can be related to the temperature. We are going to assume that the thermometer is measuring the rotation, vibration, and kinetic energy (if a liquid or gas).

How can chemical energy be measured?

The absolute amount of energy a chemical reaction contains is very difficult to measure. However, we can measure the amount of change very accurately, and in chemistry we are very interested in the change of energy. Therefore, in a chemical system, we can measure the temperature of matter very precisely. From the chapter on gases we can relate the temperature to the amount of kinetic energy the matter has. The unit of energy is the Joule (J).

How does energy transfer?

Heat is defined as the movement of energy from a higher level to a lower level. Energy will continue to be transferred until the energy of both objects is equal. To better understand the concept of heat flow, another term is introduced, the heat capacity of a substance. Heat capacity is the amount of energy that has been used to increase the temperature of 1 g of water by one degree Kelvin. The unit of heat capacity is J/g K. Remember that the change of temperature of 1 K is equivalent to 1 C. Every substance has a unique value for this physical property. The higher the heat capacity, the more energy it takes to change the substance's temperature.

What affects the transfer of heat?

As atoms and molecules vibrate and rotate, intermolecular interactions can reduce the molecules' movement. This interaction can increase the temperature needed for a phase change or the ability to transfer energy from one substance to another. There are several different types of intermolecular forces, which are listed below in terms of increasing strength of the interaction: instantaneous dipole/instantaneous dipole (also called Van der Waals forces or dispersion forces), dipole/instantaneous dipole, dipole/dipole, dipole/ion, and hydrogen bonding.

What is an instantaneous dipole/instantaneous dipole interaction?

This type of interaction occurs when the molecules of a substance contain purely covalent bonds. It is also called hydrophobic interaction, since many substances that exhibit this type of interaction are immiscible in water. The molecules are surrounded by the molecular orbitals of the atoms. The electrons are usually evenly distributed around the molecule. However, since the electrons move, there is a small probability of having an excess of negative charge and positive charge on the molecule. This excess charge is very small but noticeable. If two molecules happen to undergo this electronic transition at the same time and are next to each other, the positive and negative sides are attracted to each other, and it takes more energy for the molecules to move. If one molecule undergoes this transition, it may force the nearest molecule to undergo an electronic transition, and you end up with a slight interaction between the molecules. This interaction does not last long, but it will occur many times. Figure 13.1 is a sketch of how this will occur.

Figure 13.1 Schematic of London Dispersion Forces

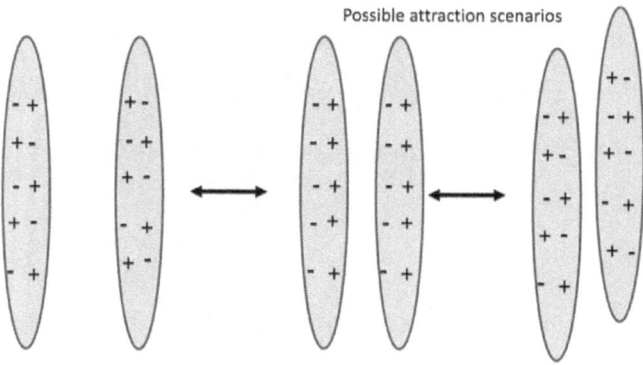

What is a dipole/instantaneous dipole interaction?

This interaction usually occurs when two different substances are mixed together. A molecular dipole was discussed in chapter 7, so when a molecular dipole is next to a hydrophobic compound, an instantaneous dipole is forced on the hydrophobic compound and the two molecules share a brief interaction. This type of interaction explains why emulsions occur and why the melting and boiling points of mixtures are different than the pure compounds.

Figure 13-2 Schematic of dipole/instantaneous dipole interaction

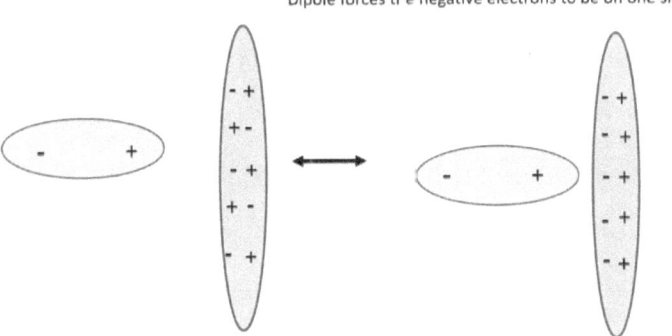

What is a dipole/dipole interaction?

This interaction is stronger than the previous two interactions. This interaction occurs when a polar molecule is attracted to another polar molecule. The interaction can be described as the slightly positive end of the molecule being attracted to the slightly negative end of the molecule. This will cause an increase in the amount of energy needed for a change in phase and to change the melting and boiling points of the mixture.

Figure 13-3: Schematic of dipole/dipole interaction

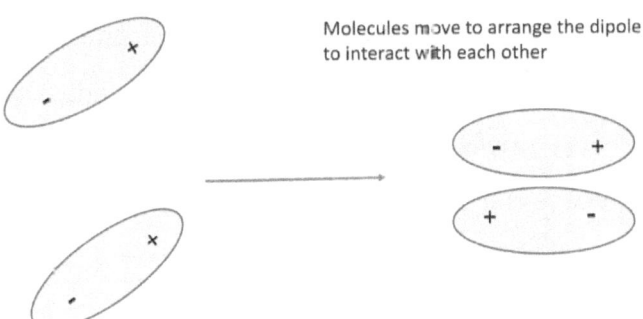

What is dipole/ion interaction?

The interaction between a dipole and an ion is the interaction that causes ionic solids to dissociate in polar solvents. The positive end of the solvent molecule will interact with the anion, breaking the ionic interaction between the cation and anion. Of course, the negative end of the dipole is interacting with the cation and breaking the interaction between the two different ions.

Figure 13.4: Schematic of dipole/ion interactions

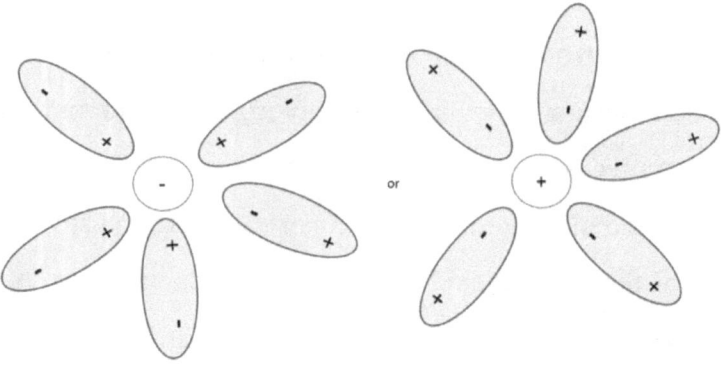

What is hydrogen bonding?

Hydrogen bonding, first discussed in chapter 1 of part two, is a unique interaction between a hydrogen and an electronegative element in a molecule. For this interaction to be present, the hydrogen must be covalently bonded to either a nitrogen, oxygen, or a sulfur atom in the molecule. The hydrogen is then attracted to a lone pair of electrons on another molecule. The atom involved in this interaction must be nitrogen, oxygen, or sulfur. This interaction is a bond between the hydrogen and the lone pair of the electronegative atom. The antibonding orbital attached to the hydrogen accepts the lone pair of the electronegative atom creating a very weak bond. This bond is relatively short lived in liquids and gases, but influences the amount of energy needed to change the phase of the substances.

Figure 13.5: Schematic of hydrogen bonding

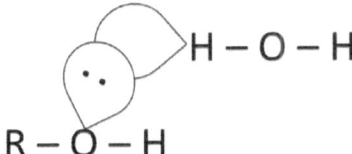

The empty antibonding orbital of hydrogen is interacting with the lone pair of oxygen

Figure 13.6: Orbital diagram of a hydrogen bond

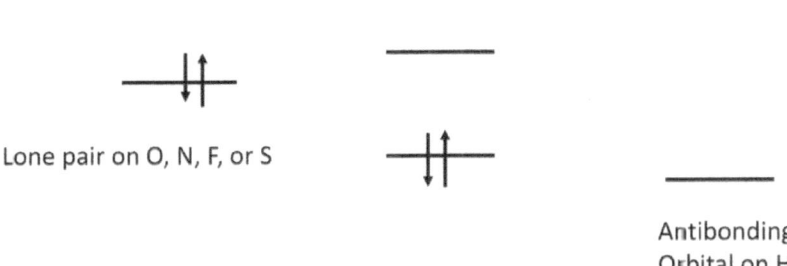

Orbital energy diagram of a hydrogen bond

How do we measure the heat of a phase change?

The amount of heat needed to change the phase of a substance depends on the type of interactions between the molecules of the substances. All molecules have London dispersion forces. The size of the molecule will dictate if the strength of the interaction can be predominating or not. The amount of energy needed to change a phase of a substance can be measured in a system called a calorimeter. This scientific apparatus consists of a double wall vessel, much like a thermos bottle, a thermometer, and a heat source. Sometimes a metal stirrer is also included, depending on the experiment.

Experimentally, calorimetry is a chemical technique used to determine the amount of heat lost or gained in a reaction. This heat, which is defined as enthalpy, is calculated from the temperature change in the reaction. In this technique, the reaction is isolated from the universe by placing the material in a special container. Before delving into the theory of thermodynamics, more terms are needed. The movement of energy is between three parts: the universe, the surroundings, and the system. The universe is everything around us and contains the surroundings and the system.

What is a system and surroundings to a chemist?

The system for a chemist is the substances that are undergoing a change in energy or a chemical reaction. This can be difficult to determine for the beginning student of chemistry. The surroundings are the environment that surrounds the system. The system and surroundings can be isolated from the universe experimentally. This can help measure the heat transfer of the chemical reaction.

As shown in Figure 13.7, the system and surroundings combination can be one of three options:

- Open system, when matter and energy can be transferred with the surroundings and system.
- Closed system, when matter is unable to move into or out of the surroundings. Energy can move between the surroundings and the system. Most chemical reactions are performed in a closed container.
- Isolated system, when both matter and energy are not able to move from the system to the surroundings and vice-versa. Precise calorimetry experiments are performed using this type of system.

Figure 13.7 Types of Thermodynamics Experiments

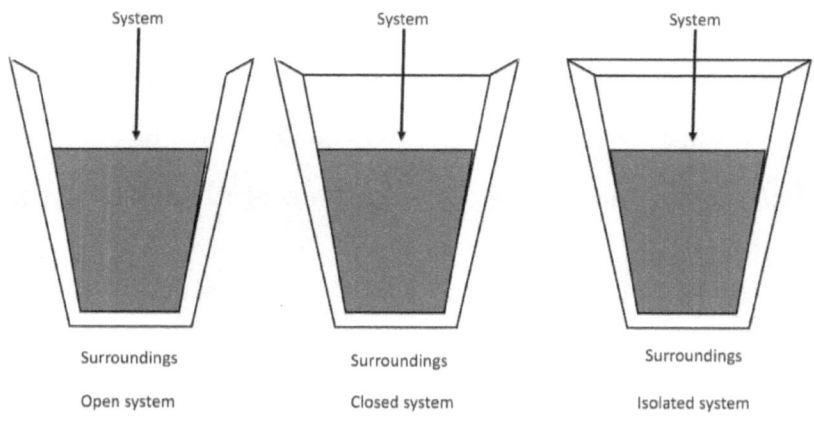

When a reaction occurs, what is important to the beginning chemist is the conditions at the start of the reaction and at the end of the reaction. Since heat transfer is a concern, the change in temperature of the reaction is needed.

What is the relationship between heat and phase changes?
If we take a solid in a calorimeter and carefully add heat at a constant rate, the temperature will increase. When the solid starts to become a liquid, the temperature will become constant until all the solid becomes a liquid. As soon as the substance becomes a liquid, the temperature will increase until the liquid starts to vaporize. At this point, the temperature will remain constant until all the liquid become a gas. A sample chart is shown in Figure 13.8.

Figure 13.8: Heat – Temperature Graph

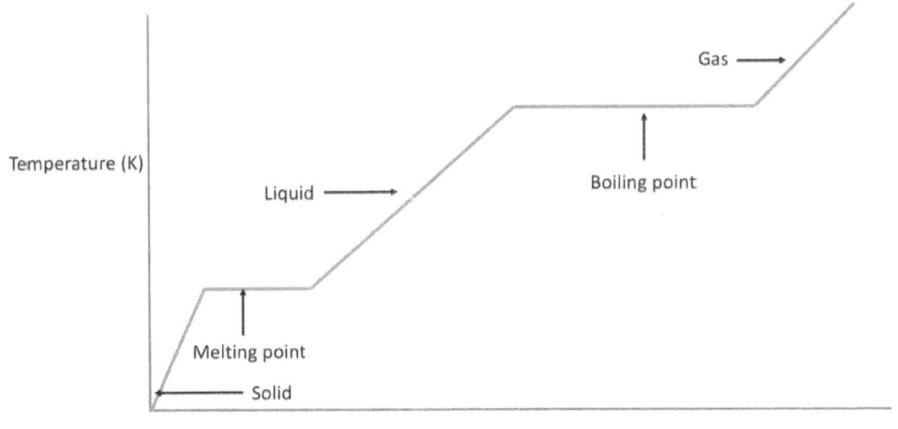

Using this chart, the parts of the line where the slope is zero is defined as the phase transition. At this point, both phases can exist at the temperature. The slope of the line will provide the inverse of the heat capacity.

Example 1:
Calculate the heat absorbed when a 50 g block of iron (C_p = 25.1 J/gK) changes its temperature from 25 °C to 50 °C. With this example, the temperature change is 25 °C, which is identical to the temperature change in Kelvin. The equation to use is listed below:

$$q = m * Cp * \Delta T$$

Where q is the heat absorbed or lost, m is the mass of the object, C_p is the heat capacity of the object and ΔT is the change in temperature of the object ($T_f - T_i$).

The heat lost is calculated to be 31,375 J or 31.4 kJ.

How can energy flows be calculated?

When chemists determine heat flow, a calorimeter is used to contain the heat so the energy transfer can be measured. The heat flow into the calorimeter can be calculated fairly easily, since the specific heat of the calorimeter is known. The relationship between the heat lost by one object to the heat gained by the other is described below:

$q_1 = -q_2$

or $m_1 * C_{p1} * \Delta T = -m_2 * C_{p2} * \Delta T$

$\Delta T = (T_f - T_i)$

The change in temperature is the change of temperature for the object, which will be different for object 1 compared to object 2. This means that one ΔT will be positive and the other negative.

Example 2

A 50g iron object with an initial temperature of 55 °C was dropped into a calorimeter containing 100 g of water at an initial temperature of 20 °C. Calculate the final temperature of the object and water.

Now, m_1 = 50g, m_2 = 100g, C_{p1}, which is iron, is 25.1 J/g°C, and Cp2, which is water, is 4.184 J/g°C. The difficulty for some students is the change in temperature. For the left side of the equation, which is iron, the change in temperature is below:

$\Delta T = (T_f - 55)$ and for water $\Delta T = (T_f - 20)$

Now to substitute all the data for the variables into the equation.

$$50\ g * 25.1 \frac{J}{g °C} * (T_f - 55) = -100\ g * 4.184 \frac{J}{g °C} * (T_f - 20)$$

Combine the data to T_f by itself.

$$\left(1255 \frac{J}{°C} * T_f\right) - 69025\ J = -\left(418.4 \frac{J}{°C} * T_f\right) + 8368\ J$$

$$1673.4 \frac{J}{°C} * T_f = 77393\ J$$

Solve for T_f, which is 46 °C. Remember, always round your answer at the end of the problem, not during the work.

Chapter 14: Thermodynamics

What are thermodynamics?

Thermodynamics is the study of the transfer of heat. For chemistry, it is the study of the transfer of energy. To study this this branch of chemistry, the scientist must keep certain variables constant or steady. Which variable is held constant will dictate which type of heat transfer to study. In this text the three types of energy being discussed are enthalpy, entropy, and Gibb's free energy, and we will discuss how to experimentally determine these values.

What is a state function?

A state function can be described as a value that is dependent only on the physical or chemical conditions or the state of the system. How the system achieved this value is immaterial to the value. This is important to a chemist, because the chemist measures the initial and final values of the system to determine the change of the system. If the path was important, the final change of energy would be different depending on the experiment.

Are there other ways to transfer energy?

In the last chapter the focus was on the transfer of heat to measure energy change. There are other mechanisms of energy transfer. In chapter 4 of part 1 of the book, the change of energy in atoms was determined by the wavelength of light emitted or absorbed. Later in chapter 18, the change of energy can be measured by chemical and electrical potential. In this chapter, work performed is used to determine energy transfer.

How is work used to measure energy change?
In physics, the work energy theorem is used to equate work to energy. Work is the force required to accelerate an object in a certain distance, which is mass times acceleration times distance or w=m*d*a. The unit of work is $kg*m^2/s^2$. Since kinetic energy is one half times mass and the square of velocity, $KE = 1/2mv^2$. The unit is also $kg*m^2/s^2$. These terms are identical but calculated by different equations and have different meanings, just the units are the same. In chemistry, work is a little harder to define.

The ideal way to explain work in a chemical sense is to look at a gas reaction in a sealed container with flexible walls. A plastic syringe is an example of a container having flexible walls. In Chapter 1 of part 2, if the temperature of the gas is increased, either the volume or the pressure must change. Since our container is flexible, the volume can change while the pressure of the gas remains constant. This change in volume, multiplied by the pressure of the gas, is the work performed on the surroundings by the gas. Therefore, use the following equation:

$$w = P\Delta V$$

where w = work, P is pressure and ΔV is the change in volume. Note that pressure and volume are considered to be state values, thus the variables are capitalized, but work is not.

What is enthalpy?

Enthalpy, ΔH, is described as the transfer of heat or energy when the pressure of the system is held constant. Using a calorimeter described in the previous chapter, a scientist can easily calculate the heat or enthalpy of a reaction. Remember, in this experiment pressure is constant, usually atmospheric pressure, and the volume can change, especially if a gas can escape the container.

What is entropy?

Entropy, ΔS, is often described as the change in disorder in the system. This disorder can be attributed to the production of a gas, formation of a solid, or what chemist call the change in the microstates of the system. Using the states of matter as an example, solids are usually highly ordered crystal structures. To place them in a solid takes energy. If the temperature of the system increases, the solid becomes a liquid and the molecules in the liquid gain movement. The molecules are rotating and vibrating as they move and the system gains disorder.

What is Gibbs' free energy?

Gibbs' free energy, ΔG, is the amount of energy change of the surroundings. To calculate the change in Gibbs' free energy, the change in enthalpy is subtracted by the change in entropy times the temperature of the system or $\Delta G = \Delta H - T\Delta S$. Later in the book there are methods to calculate ΔG without knowing ΔH or ΔS. For a reaction to be spontaneous ΔG must be negative. That means either ΔH is negative, exothermic, or ΔS is positive, gains disorder. This topic will be complicated and a college textbook will describe the differences in deeper detail.

What is internal energy of the system?

In the previous chapter, we reviewed how energy is transferred from one substance to another. As we measure the temperature change, either pressure or volume is kept constant. Enthalpy changes, ΔH, occur when pressure is constant. The change in internal energy, ΔE, occurs when volume is constant. Remember, work is done when the volume changes.

What is an exothermic or endothermic reaction?

When a chemical reaction occurs, there is a change in energy that involves kinetic, vibrational, rotational, and chemical potential energy. The energy levels of the substances involved either increase or decrease. If the total energy levels decrease, energy is given off in the form of heat and this reaction is classified as an exothermic reaction. If the energy levels increase, energy is taken in by the substances and this reaction is classified as endothermic. Another example of these terms is that if you hold the test tube that contains the reaction and it feels cold, it is an endothermic reaction. If the tube feels hot, then it is an exothermic reaction. The difficulty with these terms is the perspective of the reaction. The substances are decreasing in energy, or energy is emitted from the perspective of the substances. From your perspective, you are gaining energy because the tube feels hot in your hand. Conversely, an endothermic reaction feels cold to you because the substances need energy to react and it takes the heat from your hand.

How to calculate entropy and Gibbs' Free Energy?
The experimental calculation for determining entropy will be left for a different text. However, entropy measurements will be used to help calculate the amount of entropy change in a reaction. Gibbs' free energy will be treated identically in this chapter. Knowing how to use the results is an important part of chemistry. Tables of data can be found in any reference book or chemistry website.

What is the heat of formation?
The heat of formation is also called the enthalpy of formation. The change in enthalpy is measured for the formation of a molecule

from the elements. A sample reaction is below, the formation of NaCl:

$$Na_{(s)} + \frac{1}{2}Cl_{2(g)} \rightarrow NaCl_{(s)}$$

ΔH_f = -411 kJ/mol

By definition, each element has a heat of formation of zero. The ½ in front of the chlorine is used because the heat of formation is for 1 mole of product. If we double the moles of each compound in the equation to remove the fraction, the heat of formation is also multiplied by two. Chemists often create new compounds, and this is important because chemical engineers need to know the heat of a reaction and a reasonable estimate can be determined using Hess's law.

Table 14.1: Thermodynamic Data of Selected Elements and Compounds

Formula	Name	State	ΔH_f (kJ/mol)	ΔG_f (kJ/mol)	S (J/mol)	C_p (J/molK)
Al	Aluminum	s	0.0		28.3	24.4
AlCl$_3$	Aluminum chloride	s	-704.2	-628.8	110.7	91.8
Cl$_2$	Chlorine	g	0.0		223.1	33.9
Cu	Copper	s	0.0		33.2	24.4
H$_2$O	Water	l	-285.8	-237.1	70.0	75.3
H$_2$O	Water	g	-241.8	-228.6	188.8	33.6
C	Carbon (graphite)	s	0		5.7	8.5

CO_2	Carbon dioxide	g	-393.5	-394.4	213.8	37.1
C_2H_5OH	ethanol	l	-277.7	-174.8	160.7	112.3
C_2H_5OH	ethanol	g	-235.1	-168.5	282.7	65.4
$C_2H_4O_2$	Acetic acid	l	-484.5	-389.9	159.8	123.3
Fe	Iron	s	0.0			25.1
Fe_2O_3	Iron (III) oxide	s	-824.2	-742.2	87.4	103.9
$FeCl_3$	Iron(III)chloride	s	-399.5	-334.0	142.3	96.7
$FeCl_2$	Iron(II)chloride	s	-341.8	-302.3	118.0	76.7
Ag	Silver	s	0		42.6	25.4

C_p is the heat capacity of the substance.

Table 14.1 is republished with permission of Taylor and Francis Group LLC Books, from CRC Handbook of Chemistry and Physics, 73rd edition, published 1993; permission conveyed through Copyright Clearance Center, Inc. "

A note of warning concerning these tables. Data is different based on the source of the values and the age of the data. These values have been updated several times to reflect a more accurate or precise measurement. Do not stress over these changes when working on problems, just use data from the same source or time.

What is Hess's Law?

This law states that the energy change of a reaction that occurs in multiple steps, notably the enthalpy of the reaction, can be determined from the sum of the enthalpy of reactions of each step. This law can help chemists determine the enthalpy of a proposed reaction. Chemical engineers use this law to help design reaction vessels and chemical factories. Often, the enthalpy of the formation of a molecule is used to determine the heat of reaction. To determine the heat of reactions, sum the heat of formation for the products then subtract the heat of formation for the reactants. For example, the heat of reaction of ammonia to form hydrazine can be calculated below:

$$2\ NH_{3(g)} \rightarrow N_2H_{4(g)} + H_{2(g)}$$

$$\Delta H_r = \Sigma \Delta H_{f\ products} - \Sigma \Delta H_{f\ reactants}$$

ΔH_f of $NH_{3(g)}$ = -45.9 kJ/mol

ΔH_f of $N_2H_{4(g)}$ = 95.4 kJ/mol

ΔH_f of $H_{2(g)}$ = 0 kJ/mol

ΔH_r = (95.4 kJ/mol + 0 kJ/mol) – (2*-45.9 kJ/mol) = 187.2 kJ/mol

The heat of the reaction is positive, meaning that this is an endothermic reaction and heat is needed. The heat of formation of NH_3 is multiplied by two because of the coefficient in the balanced equation.

Chapter 15: Kinetics

Kinetics is the study of the speed of a reaction and how the reaction takes place. This branch of chemistry is very important to organic and medicinal scientists. Some reactions are favored to occur as written, but they may take a long time for the reaction to go to completion. Many geological processes are chemical in nature but since it takes a long time for the products to form, it is said to be a kinetically slow reaction. On the other hand, many reactions are almost instantaneous or very quick. In this chapter, we will discuss the various types of kinetic reactions and solve simple first order reactions.

What is meant by reaction rate?

The reaction rate is how fast the reaction will occur under the specified conditions. If the conditions are changed, the rate may be changed. To measure the reaction rate, the formation of a product or the loss of a reagent is measured over a period of time. The rate is negative if the amount of reactant is decreasing and the rate is positive if the product is measured. The unit for rate is molarity per second. The equation to calculate the rate is the following: $rate = \frac{\Delta[x]}{\Delta t}$, where $\Delta[x]$ is the change in concentration and Δt is the change in time.

How is the rate measured?

Reaction rates are measured by taking concentration measurements over time. Sometimes the decrease of a reagent is easy to measure. If this is the case, the rate is negative since the loss of material is measured. If the formation of a product is measured, the rate is positive. Therefore, most kinetic data describes what compound is being measured.

What conditions can affect the reaction rate?

There are several factors that affect how fast a reaction will occur. Temperature, concentration of a substance, surface area of a substance, presence of a catalyst or an inhibitor can affect the rate of reaction.

How does surface area change the rate of reaction?

The size of the particle is very important for a reaction that involves a solid. Whether it is a solid reacting with a gas, a solid reacting with a liquid, or a solid as a catalyst or inhibitor, the size of the solid particle is significant. Bigger is not better in this case. In fact, the smaller the particle, the faster the reaction. It has to do with the surface area of the particle. If you have a solid particle with a dimension of 1 cm on each side, you have a surface area of 6 cm². Now take that particle and divide it so that the dimensions are 0.1 cm on each side. The surface area of 1

51

particle is 0.6 cm², but now we have 1000 particles and the total surface area is 600 cm². Therefore, we can have more reagents reacting with the smaller particles than the larger ones.

How does a catalyst/inhibitor affect the reaction rate?

A catalyst is a substance that lowers the activation energy of a reaction and helps to increase the reaction rate. This substance can be an inorganic or organic compound. Often used in biochemistry, organic and inorganic chemists use catalysts to help speed up the reaction. The catalyst is not consumed in the reaction. To identify the presence of the catalyst, the substance is placed above or below the reaction arrow. See the decomposition of hydrogen peroxide using manganese dioxide as a catalyst:

$$2\ H_2O_2 \xrightarrow{MnO_2} 2\ H_2O + O_2$$

An inhibitor is similar to a catalyst; however, it slows a reaction down by increasing the activation energy needed to complete the reaction.

How does temperature affect kinetics?

The temperature of the reaction is dependent on whether the reaction is exothermic or endothermic. Given an exothermic reaction, the temperature of the surroundings needs to be lower than the system. This ensures the reaction will go to completion. If the temperature is higher than the system, it will act as an inhibitor and slow the reaction down. For example, the disassociation of sodium hydroxide will be faster if the flask is placed in cold water. For an endothermic reaction, the temperature of the surroundings will need to be hotter than the system to speed it up. For example, the dissolution of sugar will increase in rate if the temperature of the water is higher than the surroundings.

Is there an equation describing the temperature effects?
Arrhenius developed an equation that describes the effect of temperature on a reaction. For most reactions, the relationship between rate and temperature is found in the rate constant and dependent on the natural logarithm of temperature and the activation energy:

$$k = Ae^{-Ea/RT}$$

Where A is called the frequency factor, E_a is the activation energy, T is temperature in Kelvin, and R is the gas constant in units of J/mol*K

How does concentration affect the rate of reaction?
Concentration affects the reaction differently depending on the reaction order of the compound in the reaction. Most of the time an increase in the concentration of a reactant will increase the rate of the reaction. However, the factor of increase will be different.

What is the rate law?

The rate law is an equation that explains the effect of concentration on the rate of the reaction. It is comprised of a rate that is equal to a constant and the concentration of the reagents raised to a power. The power indicates how many molecules are needed for the reaction. However, this is not the coefficient of the substance in the chemical equation. It is experimentally determined and it explains the how the reaction can occur or the mechanism of the reaction.

What is the rate constant?

The rate constant is a constant that is unique for each reaction and it is used to determine how long the reaction will occur. What is special about rates and rate constant is the rate constant can provide the chemist clues on the reaction pathway. Many reactions have more than one step to make the products, and one of these steps is the slowest. Finding this step is important to chemists and chemical engineers. Also, chemical engineers use rate constants to determine how long a mixture should be in a reaction vessel to make the amount of product.

What is a first order reaction?

A first order reaction is a reaction that relies on one reagent being involved in the reaction, or one reagent setting up the rate-limiting step. Many decomposition reactions are a first order reaction. The equation for a first order reaction is:

$$ln\left(\frac{[x]_t}{[x]_0}\right) = -kt$$

Where ln is the natural logarithm, $[x]_t$ is the concentration of a compound at time t, $[x]_0$ is the initial concentration, k is the rate constant, and t is time.

There are three common types of reactions that will exhibit a first order reaction: radioactive decay, combination, and decomposition reactions. With kinetics some of the reaction will exhibit other order of reactions.

What is the actual formula for half-life?

In chapter 3 of part 1, radioactive decay was discussed and a brief description on half-lives was presented. Kinetically, a radioactive decay is a first order reaction. The reagent decay into another element and the half-life equation is identical to the first order reaction.

What is a decomposition reaction?

A decomposition reaction is a reaction involving one reagent that breaks apart into two separate entities. Some compounds will decompose to two different elements while others will break apart to a compound and an element or two separate elements. Energy is needed to start the reaction. The generic equation is written as A → B + C and an example is described below:

$$2\ HgO_{(s)} \rightarrow 2\ Hg_{(l)} + O_{2(g)}$$

What is a combination reaction?

This reaction consists of two substances and the product is a single compound. Generally, energy of some form is added to the reagents to start the reaction. Although some reaction will occur spontaneously, without added energy or influence. The reagents do not necessarily have to be elements; the important characteristic is that the reaction has two reagents and one product. Generically, it is written as A + B → C.

$$2\ H_{2(g)} + O_{2(g)} \rightarrow 2\ H_2O_{(g)}$$

Many of these reactions have a rate order of one or two, depending on the reaction conditions.

Chapter 16: Equilibrium

What is equilibrium?

In chemistry, a common type of reaction is called equilibrium reaction. This reaction may encompass a single or double displacement reaction or a decomposition reaction. In simple terms, the reaction can proceed from reaction to products and the products can combine to form the reactants. In general terms:

$$aA + bB \leftrightarrow cC + dD$$

For a system to be in equilibrium, the compounds involved in the reaction must be at a steady concentration for an indefinite period. There are also physical processes that can be in a state of equilibrium. The dissolving of gases into a liquid is one example. There are two definitions of equilibrium in chemistry, one based on kinetics and one based on thermodynamics.

What is the equation for the kinetic expression?

For a reaction below, there is a forward rate of the reaction, k_1, and the corresponding rate of the reverse reaction, k_2.

$$aA + bB \underset{k_1}{\overset{k_2}{\leftrightarrows}} cC + dD$$

The kinetic definition is that the equilibrium constant is equal to the rate of the forward reaction divided by the rate of the reverse reaction.

$$K = \frac{k_2}{k_1}$$

However, determining this value experimentally is time consuming and there is a quicker method to calculate the equilibrium constant.

What is the relationship based on concentration?

The best-known definition is based on thermodynamics, which is the concentration of the products divided by the concentrations of the reactants. Each concentration is raised by the coefficient in the balanced chemical reaction. For the general reaction given above, the equilibrium constant is below.

$$K = \frac{[C]^c[D]^d}{[A]^a[B]^b}$$

K is a unitless number, so the concentration must be divided by a normalizing factor to remove the unit before raising to the power. For dissolved solutes, the standard value is 1 M, and for gases, 1 atm. For solids and pure liquids, the value of the concentration is unity, and in the calculation the pure solid and liquids have no bearing on the final answer. Gases are considered in a mixture because most reactions are performed in the open air where there are mixtures of gases. Here, this relationship will be used extensively because of the important power that this relationship has in solution chemistry. Since concentration affects kinetics, there is a relationship between the two definitions of equilibrium. However, the mathematical relationship between the kinetic expression and the thermodynamic expression is left for a college level text book.

How can a gas be in equilibrium with a liquid?

A simple, easy equilibrium relationship to study is a gas being dissolved into a liquid. All gases can be dissolved into a liquid. Oxygen dissolving into water is very important to sea life. The amount of gas in the liquid is based on the solvent, and the equilibrium constant for a gas dissolved in a liquid is unique for each gas/liquid pair. This relationship is called Henry's Law. For this interaction, we can write a chemical equation for the dissolving of a gas into a liquid using the relationship below.

$$X_{(g)} \leftrightarrow X_{(liquid)}$$

The equilibrium constant for this generic chemical equation is:

$$K_H = \frac{[X]}{P_X} \text{ or } P_X K_H = [X]$$

Where P_X is the pressure of the gas above the surface of the liquid, $[X]$ is the concentration of the gas in the liquid, and K_H is the equilibrium constant – called Henry's Law constant. With this equilibrium, the pressure of the gas forces the gas into the liquid until the liquid is saturated. Cooling the liquid will enable more gas to be dissolved into the liquid, this is a reason why most soft drinks are served cold, to keep the gas dissolved in the liquid. A liquid being dissolved in the air is slightly different due to the difference in phase characteristics. The amount of the liquid being dissolved in the gas is dependent on the temperature of the system.

How is equilibrium determined in gas reactions?

For reactions in a gas, the relationship between pressure and concentration is needed. While concentration in liquids is an easy concept to describe, the concentration of a gas is a little different concept. Let's begin with a container that contains a vacuum or nothing. There is no matter but a volume of nothing. Add a little gas into the container and now we have introduced a substance in a volume and the concentration of the gas, in moles per liter, can be calculated. However, if we use the common terms for concentration, we can confuse students and other scientists. Now, using ideal gas laws, we know the relationship between pressure and volume of a gas:

$$P = \frac{nRT}{V}$$

Where P is pressure, n is the number of moles in the system, R is the gas constant, T is temperature and V is volume. Now, n/V is also considered to be concentration, where V is the volume of container – not the solvent. This is important, even if the solvent is a gas. We are still looking at the container for the volume. Now we have a relationship between pressure and concentration. If we have more than one compound present in the gas phase, the pressure becomes the partial pressure of the compound. For the equilibrium expression, we can replace the concentration terms with partial pressure of the gases. For example, the decomposition of $N_2O_{5(g)}$:

$$N_2O_{5(g)} \leftrightarrow NO_{3(g)} + NO_{2(g)}$$

$$K = \frac{P_{NO_3(g)} P_{NO_2(g)}}{P_{N_2O_5(g)}}$$

How is equilibrium determined in reactions in solutions?

Reactions in solutions can be very simple or complicated, depending on the chemistry involved. For this section of the book, the focus will be on the following processes: double displacement reaction, dissolution of ionic compound, and the dissolving of a molecular compound.

The dissolving of a molecular compound

Another term for this concept is solubility and it was discussed in chapter 11. Like dissolving of a gas in a liquid, solid dissolving in a gas is a simple concept. Realize that the units are different between solubility values and equilibrium constant. To convert the solubility value to an equilibrium value, the mass of the solvent must be converted to liters and the mass of solute must be converted to moles. The equilibrium constant expression for the example of dissolving sugar in water is described below:

$$sugar_{(s)} \leftrightarrow sugar_{(aq)}$$

$$K = [sugar_{(aq)}]$$

K is the solubility constant and brackets indicate the concentration of the solute.

Is there a difference between solubility and the solubility constant?

Solubility is the amount of a compound or an ion that is soluble in 100 mL of solvent. Remember, the solubility of sugar and the solubility of sodium chloride is mass per 100 mL. The solubility product is a constant based to the maximum concentration of each ion.

The dissolution of an ionic solid

The dissolution of an ionic solid is a chemical property where the water breaks up the crystal structure of an ionic solid to form the cation and anion in solution. Each ionic solid has a unique solubility product in water (K_{sp}), which is a term for the equilibrium constant for the dissociation of a solute in water. The smaller the K_{sp}, the lower the solubility of the compound. This constant will never be zero, but it will close to it. Water's unique property helps to keep many compounds slightly soluble. Using the dissociation of $CaCl_2$ as an example, the chemical equation and equilibrium expression is described below:

$$CaCl_{2(s)} \overset{H_2O}{\leftrightarrow} Ca^{2+} + 2\,Cl^-$$
$$K = [Ca^{2+}][Cl^-]^2$$

For this text, since the solvent is water, the aqueous symbol is missing in the products, since the products are ions. Also, the solid is not included with the equilibrium expression since the value of a solid is 1.

What are the Solubility Rules?

To determine if an inorganic compound will be highly or slightly soluble in water, the following rules are a good guide to determine relative solubility.

1. All salts containing an alkali metal (Li, Na, K, Cs, and Fr) will be soluble in water.
    a. Hydrogen ion will also be soluble, but this cation is special and will be discussed later in the chapter.
2. All salts containing ammonium ion (NH_4^+) will be soluble in water.
3. All salts containing nitrate (NO_3^-) will be soluble in water.
4. All salts containing a halide (Cl^-, F^-, Br^-, I^-) will be soluble in water – except
    a. Except for Ag^+, Pb^{2+}, Hg_2^{2+}
    b. Hydride anion (H^-) will be soluble and reactive in water and will be discussed in other text books.
5. All salts containing hydroxide anion (OH^-) will be insoluble
    a. Except for the alkali metals and Ca and Ba
6. All salts containing sulfides (S^{-2}) are insoluble
    a. Except for alkali and alkali earth metals

There are always exceptions to the rules, especially when the working with organic compounds.

Are insoluble salts soluble in water?

Many compounds that are considered insoluble by many people are in fact slightly soluble, especially in water. Water has this unique ability to dissolve many organic compounds and inorganic salts. For the solubility of an inorganic salt, the dissolution reaction is written first, for example, the dissociation of sodium chloride:

$NaCl_{(s)} + H_2O \rightarrow Na^+ + Cl^-$

Then an equilibrium constant can be calculated:

$K_{sp} = [Na^+][Cl^-]$

K_{sp} means the equilibrium constant for the solubility product, which is the concentration of the cation times the concentration of the anion. If the salt has more than two ions because of the dissociation, then the K_{sp} relationship will reflect the multiple parts by taking the coefficient and using it as the power of the concentration. Below is a small table of K_{sp}, solubility product constants.

Table 16.1: Solubility Product Constants

Substance	formula	K_{sp}
Barium Carbonate	$BaCO_3$	2.58×10^{-9}
Calcium carbonate	$CaCO_3$	4.96×10^{-9}
Copper (I) chloride	$CuCl$	1.72×10^{-7}

Iron (III) hydroxide	Fe(CH)$_3$	2.64×10^{-39}
Lead iodide	PbI$_2$	8.49×10^{-9}
Mercury (I) iodide	Hg$_2$$_2$	5.33×10^{-29}
Silver chloride	AgCl	1.77×10^{-10}
Silver iodide	AgI	8.51×10^{-17}
Tin (II) sulfide	SnS	3.25×10^{-28}
Nickel (II) sulfide	NiS	1.07×10^{-22}
Tin(II) hydroxide	Sn(CH)$_2$	5.45×10^{-27}

Table 16.1 is republished with permission of Taylor and Francis Group LLC Books, from CRC Handbook of Chemistry and Physics, 73rd edition, published 1993; permission conveyed through Copyright Clearance Center, Inc. "

For example, the dissociation of calcium hydroxide.

$$Ca(OH)_{2(s)} + H_2O \rightarrow Ca^{2+} + 2\ OH^-$$

Now the K$_{sp}$ relationship is shown below:

$$K_{sp} = [Ca^{2+}][OH^-]^2$$

Each compound has a unique value for the K$_{sp}$, and the smaller the K$_{sp}$, the less soluble the salt is in the solvent. As an aside, all ionic compounds can be dissolved in organic solvents to a degree and polar organic solvents can dissociate the salts to a degree. To calculate the concentration of each ion in pure water, the K$_{sp}$ is equal to the concentration of each ion raised by the coefficient.

Example 1:

Calculate the concentration of each ion in the dissociation of tin hydroxide.

The reaction is:

$$Sn(OH)_2 \xrightleftharpoons{H_2O} Sn^{2+}_{(aq)} + 2\,(OH)^-_{(aq)}$$

and the equilibrium expression is:

$$K_{sp} = [Sn^{2+}][OH^-]^2$$

Algebraically, for every Sn^{2+} cation, there are two OH^- produced. The expression is if $x = Sn^{2+}$, then $2x = OH^-$. Substitute into the K_{sp} expression with the value for the K_{sp}.

$5.45 \times 10^{-27} = x(2x)^2$. Now solve for x

The $[Sn^{2+}] = 1.11 \times 10^{-9}$ M and $[OH^-] = 2.22 \times 10^{-9}$ M

The double displacement reaction.

This reaction is one of the classical class of reactions. Two solutions, each containing a salt dissociated in water, are added together and solid forms. This reaction combines the concepts of ionic solubility and the dissociation of ionic compounds. The products from this reaction are either a solid, a gas, or a liquid.

What causes the formation of a solid?

When two solutions containing inorganic salts are mixed together, there is a possibility that a solid, gas, or liquid is formed. When a solid is formed, a cation and anion are attracted to each other and overcome the electrostatic attraction of water to form a solid. The attraction is based on several hypotheses including the charge on the ion, the hard or softness of the ion, and size of the anion. For this text, a general description of the results of the reactions will be discussed, the why will be described in a college text.

How does this relate to the formation of the solid?

If two solutions containing soluble salts are mixed together a reaction occurs, which means a solid, liquid, or a gas is formed. To produce a solid, the resulting solid is highly insoluble in water. For example, look at the reaction of KI with $Pb(NO_3)_2$:

$$2\ KI_{(aq)} + Pb\ (NO_3)_{2(aq)} \rightarrow PbI_{2(s)} + 2\ KNO_{3\ (aq)}$$

The equilibrium expression is:

$$K_c = \frac{[KNO_3]^2}{[KI]^2[Pb(NO_3)_2]}$$

But, if we reduce the balanced chemical reaction to the balanced ionic equation

$$2\ K^+ + 2\ I^- + Pb^{2+} + 2\ NO_3^- \rightarrow PbI_{2(s)} + 2\ K^+ + 2\ NO_3^-$$

and

$$2\ I^- + Pb^{2+} \rightarrow PbI_{2(s)}$$

The equilibrium expression becomes:

$$K_c = \frac{1}{[I^-]^2[Pb^{2+}]} = K_{sp}^{-1}$$

So knowing the ionic equation helps to simplify the equilibrium expression.

What happens when there are additional cations or anions in the solution?

In all of the examples above, the solution contains no other ions. So what happens when the solution contains additional cations or anions. The amount dissociated will change and the ionic strength of the solution will change, affecting the solubility of the compound. For an excess of the ion being dissolved in solution, the compound will not dissociate as much. The example below will illustrate this concept.

Example 2:

Solid silver iodide is added to a 1.0 M solution of sodium iodide and to pure water, what is the concentration of silver present in each solution. For a pure solution of water, the concentration of silver is determine by solving the following equation:

$K_{sp} = 8.57x10^{-17} = [Ag^+][I^-]$

$x * x = 8.57x10^{-17}$

$x = [Ag] = 9.26x10^{-9} M$

Now for the 1.0 M solution of sodium iodide

$K_{sp} = 8.57x10^{-17} = [Ag^+][I^-]: \quad [I^-] = 1.0\ M$

$x * 1 = 8.57x10^{-17}$

$x = [Ag] = 8.57x10^{-17} M$

There is less silver dissolved in the 1.0 M sodium iodide solution, because the excess iodide suppresses the dissociation of the silver iodide. Now, if the solution contains a compound which is spectator ions when dissociated in water, the ionic strength of the solution increases and the overall dissociation of the target compound will increase. The theoretical calculations will be left for a college level textbook.

Chapter 17: Reactions Involving Acids and Bases

What are acid/base reactions?

This is a different classification of a double displacement reaction. Acid/Base reactions or neutralization reactions, this class of reactions have a common dominator of forming water and a salt when an acid and base are mixed together. Depending on the reagents, the salt may precipitate based on the solubility rules listed in the previous chapter. For this chapter, the reagents are in an aqueous solution.

How does formation of water occur?

There is a special type of double displacement reaction in which the product is water and usually a soluble salt. For one compound in solution, the cation is hydronium ion, H^+, the other compound in solution must contain the hydroxide anion, OH^-. The hydrogen containing compound is classified as an acid and the hydroxide containing compound is called the base. When solutions containing an acid and base are mixed together, water and a soluble salt is usually formed. There are several definitions of acids and bases, depending on the chemistry that is occurring. For this text, I will focus on two acid/base systems – the Bronsted/Lowry and the Lewis System.

What is a Bronsted/Lowry Acid or Base?

The Bronsted/Lowry definition explains an acid as a compound that produces a hydronium ion (H⁺ or H₃O⁺) when dissolved in water and a base is a compound that produces a hydroxide ion when dissolved in water. This definition is useful in explaining acid/base concepts to the beginning chemist and is the most common definition used in the chemistry field For the acid, an example is the dissociation of hydrochloride in water to form hydrochloric acid.

$$HCl + H_2O \rightarrow H_3O^+ + OH^-$$

The typical example of a base is the dissociation of sodium hydroxide in water:

$$NaOH + H_2O \rightarrow Na^+_{(aq)} + OH^-_{(aq)}$$

Also, for organic acids, those compounds that have -CO₂H group in the chemical formula, the hydrogen attached to the oxygen will also dissociate. This is a simplistic explanation for organic compounds, your teacher can expand on this if you ask.

$$CH_3CH_2CO_2H_{(aq)} \longrightarrow CH_3CH_2CO^-_{2(aq)} + H^+_{(aq)}$$

You will notice that the hydrogen that disscciates is the one bonded to the oxygen, none of the others are able to dissociate. This is due to the electronegativity of the oxygen atom being higher than hydrogen and the bond is very polar, able to break easily.

For organic bases, mainly those compounds that contain a nitrogen in the compound, the reaction will remove a hydrogen from oxygen to make a hydroxide anion.

$$CH_3CH_2NH_{2(aq)} + H_2O_{(l)} \longrightarrow CH_3CH_2NH_{3(aq)}^+ + OH_{(aq)}^-$$

Not only is this reaction is classified as a Bronsted Base, it also can be classified as a Lewis Base.

What is a Lewis Acid/Base?

The Lewis acid and base definition explains an acid as a compound that accepts an electron pair when interacting or bonding and a base is a compound that donates an electron pair. This helps explains the chemistry behind hydrogen bonding and the chelation of molecules and ions to a metal center, called coordination chemistry. This field of chemistry helps explains why many inorganic salts have various colors and intensities. The example below shows how Al^{3+} can be classified as a Lewis acid:

$$Al^{3+}_{(aq)} + H_2O_{(aq)} \longrightarrow Al(OH)_{(aq)}^- + OH_{(aq)}^-$$

As shown above, water acts as a base, donating the electron pair to aluminum to form the complex and the proton remains in solution. Another example is reaction of nickel with nitrate:

$$Ni^{2+}_{(aq)} + 4\, NO_{3(aq)}^- \longrightarrow Ni(NO_3)_{4(aq)}^{2-}$$

Ni^{2+} is the acid, which accepts the electron pair from NO_3^-.

What is an acid and base reaction?

The acid compound, when added to water, will dissociate to form the hydronium cation and the corresponding conjugate base. A generic reaction equation looks like this:

$$HA \rightarrow H^+ + A^-$$

Where HA is the acid, H^+ is the proton and A^- is the conjugate base. The word conjugate in chemistry always refers to the product that is formed. A basic compound, when dissolved in water, will dissociate in water to form the hydroxide anion and the conjugate anion. Now, usually, the conjugate base is negatively charged, and the conjugate acid is positively charged. Many times, this is not the case. For example:

$$A^- + H_2O \rightarrow AH + OH^-$$

HA is the conjugate acid to the base A^-.

This can be confusing. The reverse reaction in the equilibrium is not the hydronium reacting with the conjugate base, but the conjugate base reacting with water. This is true because the amount of water surrounding the anion is significantly more than the amount of H^+ in solution.

For the dissociation of a base:

$$B + H_2O \rightarrow BH^+ + OH^-$$

the reverse reaction is:

$$BH^+ + H_2O \rightarrow B + H_3O^+$$

Again, this occurs because the number of water molecules is significantly higher than the number of hydroxide anions in solution.

What is a strong acid or base?

A strong acid is a compound that will dissociate completely to form H+ and the conjugate base. The reverse reaction will not occur, because the conjugate base is too weak to react with water to form the acid. The strong base is very similar to a strong acid; the compound will dissociate completely to form the hydroxide anion and the conjugate acid. The conjugate acid is too weak to remove the hydroxide anion from water.

What is a weak acid and base?

A weak acid is a compound that will dissociate slightly, so there is a significant amount of the acid and conjugate base in solution. An equilibrium will be reached when the forward reaction rate is equivalent to the reverse reaction rate. Since the concentration of each species will be easier to solve, an equilibrium constant can be calculated using the following equation:

$$HA \rightarrow H^+ + A^-$$

$$K_a = \frac{products}{reactants} = \frac{[H^+][A^-]}{[HA]}$$

Where K_a is called the acid dissociation constant and each K_a is unique for each acid compound.

The weak base is a compound that will dissociate slightly, so there is a significant concentration of the base and the conjugate acid in solution. Equilibrium will be reached when the forward reaction rate will be equivalent to the reverse reaction rate.

$$B + H_2O \rightarrow BH^+ + OH^-$$

$$K_b = \frac{products}{reactants} = \frac{[BH^+][OH^-]}{[B]}$$

Where K_b is defined as the base hydrolysis constant and is unique for each base. Water is a liquid and has a value of 1 for this equilibrium expression.

Is there a relationship between K_a and K_b?

In weak acid solution there will be two competing reactions, the dissociation of the acid and the hydrolysis of water according to the generic equation below:

AH + H_2O → H_3O^+ + A^-

A^- + H_2O → HA + OH^-

If you add these equations together, you will receive:

2 H_2O → H_3O^+ + OH^-

which is the autoprotolysis of water.

You can conclude that $K_a * K_b = K_w$, because if you add two or more reactions together, you will multiply the equilibrium constants. K_w is a constant and the value is 1.0×10^{-14}. This means for ultra-pure water, the concentration of H^+ and OH^- is 1.0×10^{-7} M. This relationship is used by chemistry to calculate the K_b values for bases from data tables. Below are the Ka values for examples of a weak acid and base. The step in the tables indicate which proton dissociates from the acid. Many weak acids and bases can have two dissociation steps. For example, the dissociation of carbonic acid is shown below:

Step 1: $H_2CO_{3(aq)} \rightarrow H^+ + HCO_3^-$

Step 2: $HCO_{3(aq)}^- \rightarrow H^+ + CO_3^{-2}$

Table 17.1 Examples of weak organic and inorganic acids and corresponding K_a

Name	formula	step	K_a	pK_a
Acetic acid	CH_3CO_2H	1	1.76×10^{-5}	4.75
Benzoic acid	$C_6H_5CO_2H$	1	6.46×10^{-5}	4.19
Carbonic acid	H_2CO_3	1	4.3×10^{-7}	6.37
	H_2CO_3	2	5.61×10^{-11}	10.25
Chloroacetic acid	CH_2ClCO_2H	1	1.40×10^{-3}	2.85
Formic acid	HCO_2H	1	1.77×10^{-4}	3.75
Glutaric acid	$C_5H_8O_4$	1	4.58×10^{-5}	4.31
		2	3.89×10^{-6}	5.41
Octanoic acid	$C_8H_{16}O_2$	1	1.09×10^{-5}	4.96
Pentanoic acid	$C_5H_{10}O_2$	1	1.70×10^{-5}	4.77
Phosphoric acid	H_3PO_4	1	7.52×10^{-3}	2.12
	H_3PO_4	2	6.23×10^{-8}	7.21
	H_3PO_4	3	2.2×10^{-12}	12.67
Sulfuric acid	H_2SO_4	1	strong acid	
	H_2SO_4	2	1.2×10^{-2}	1.92

Table 17.1 is republished with permission of Taylor and Francis Group LLC Books, from CRC Handbook of Chemistry and Physics, 73rd edition, published 1993; permission conveyed through Copyright Clearance Center, Inc. "

Table 17.2 Examples of weak organic bases and the corresponding K_a

Name	step	K_a	pK_a
Aniline	1	2.34×10^{-5}	4.63
Arginine	1	1.51×10^{-2}	1.82
	2	1.01×10^{-9}	8.99
Benzylamine	1	4.67×10^{-10}	9.33
2-aminoethanol	1	3.16×10^{-10}	9.50
glycine	1	4.46×10^{-3}	2.35
	2	1.68×10^{-10}	9.78
Methylamine	1	2.70×10^{-11}	10.657
3-aminopentane	1	2.57×10^{-11}	10.59

Table 17.2 is republished with permission of Taylor and Francis Group LLC Books, from CRC Handbook of Chemistry and Physics, 73rd edition, published 1993; permission conveyed through Copyright Clearance Center, Inc. "

The reaction for the compounds in Table 17.3 is the reverse of the base

$$BH^+ + H_2O \rightarrow B + H_3O^+$$

To determine the K_b of the reaction below, use the following relationship:

$$B + H_2O \rightarrow BH^+ + OH^-$$

$K_b = K_w/K_a$ or $K_b = 1 \times 10^{-14}/K_a$

What does pH mean?

Since water will dissociate by itself when an acid or base is added to water, the concentration of both H^+ and OH^- are smaller than 1.

Therefore, a system to describe the amount of H⁺ was developed to describe the concentration simply. The system developed was to take the $-\log_{10}[H^+]$. The pH scale ranges typically from 1 – 14. If the concentration of H⁺ increases the concentration of OH⁻ must decrease per the Kw relationship.

Why doesn't water have a pH of 7?

Ultra-pure water, bottle under nitrogen gas or coming out of the purification system has a pH of 7. However, the longer the water is in contact with air and especially carbon dioxide, the gas is dissolved into the water and carbon dioxide reacts with water to form carbonic acid.

$$CO_{2(g)} + H_2O \rightarrow H_2CO_{3(aq)}$$

Carbonic acid is one of few compounds that cannot be isolated chemically or physically. We infer the structure from many experiments and observations. The fact that carbonic acid can dissociate to form carbonate and hydrogen carbonate is good evidence for the presence of carbonic acid. This dissociation of carbonic acid is the reason why natural water and water received from the tap is slightly acidic. Acid rain is rain that has a pH less than 4 and contains nitric and sulfuric acid which can harm plants, animals and buildings.

Example 1

Calculate the pH of a 0.30 M solution of hydrochloric acid

$$pH = -\log[H^+]$$
$$pH = -\log(0.30)$$
pH = 0.523

Example 2

Calculate the pH of a 0.04 M solution of lithium hydroxide

$$[H^+][OH^-] = 1x10^{-14}$$

$$[H^+] = \frac{1x10^{-14}}{0.04}$$

$$[H^+] = 2.5x10^{-13} \, M$$

pH = 12.60

Below is an alternate method using pOH

$$pH + pOH = 14$$
$$pOH = -\log[OH^-]$$

pOH = 1.40

$$pH = 14 - pOH$$

pH = 12.60

Example 3

Determine the Kb of benzylamine

There are two methods to determine the Kb

$$K_a * K_b = 1x10^{-14}$$

$$K_b = \frac{1x10^{-14}}{K_a}$$

$$K_b = \frac{1x10^{-14}}{4.67x10^{-10}}$$

$$K_b = 2.14x10^{-5}$$

Second method

$$pK_a + pK_b = 14$$

$$pK_b = 14 - pK_a$$

$$pK_b = 14 - 9.33$$

$$pK_b = 4.67$$

$$K_b = 10^{-4.67}$$

$$K_b = 2.14x10^{-5}$$

Example 4

Determine the pH of a 0.025 M solution of acetic acid

Ka = 1.8x10-5

$$K_a = \frac{[H^+][A^-]}{[HA]}$$

For this problem, $[H^+] = [A^-] = X$ and $[HA] = 0.025 - X$, so

$$K_a = 1.8x10^{-5} = \frac{X*X}{0.025-X} = \frac{X^2}{0.025-X}$$

Solve for X using the quadratic equation

$$1.8x10^{-5}(0.025 - X) = X^2$$

$$X^2 = 1.8x10^{-5}X - 4.625x10^{-7}$$

$$X = \frac{-b \pm \sqrt{b^2 - 4ac}}{2a}$$

where a = 1, b = 1.85x10⁻⁵ and c = - 4.625x10⁻⁷

$$X = \frac{-1.85 \times 10^{-5} \pm \sqrt{(1.85 \times 10^{-5})^2 - 4(1)(-4.625 \times 10^{-7})}}{2(1)}$$

$X = 6.71 x 10^{-4}\ M, which\ is\ H^+$

$pH = 3.17$

Example 5

Determine the pH of a 0.048 M solution of methylamine

Similar to example 4, but use K_b instead.

$K_a * K_b = 1 x 10^{-14}$

$K_b = 3.70 x 10^{-4}$

$K_b = \frac{[BH^+][OH^-]}{[B]}$

For this problem, [BH⁺] = [OH⁻] = X and [B] = 0.048 – X

$K_b = 3.70 x 10^{-4} = \frac{X * X}{0.048 - X} = \frac{X^2}{0.048 - X}$

Solve for X using the quadratic equation

$3.70 x 10^{-4} (0.048 - X) = X^2$

$X^2 = 3.70 x 10^{-4} X - 1.776 x 10^{-5}$

$X = \frac{-b \pm \sqrt{b^2 - 4ac}}{2a}$ where a = 1, b = 3.70x10⁻⁴ and c = - 1.776x10⁻⁵

$$X = \frac{-3.70 \times 10^{-4} \pm \sqrt{(3.70 \times 10^{-4})^2 - 4(1)(-1.776 \times 10^{-5})}}{2(1)}$$

$X = 4.03 x 10^{-3}\ M, which\ is\ OH^-$

$So, [H^+] = \frac{1 x 10^{-14}}{4.03 x 10^{-3}} = 2.48 x 10^{-12}\ M$

$pH = 11.61$

What is a buffer?

A buffer is a special mixture of a weak acid and the salt of the weak acid. A buffer solution can withstand the change in pH when an acid or base is added. The chemistry of a buffer is the combination of the weak acid and weak base chemistry. For a buffer to be effective, the weak acid and base are the conjugate of each other. Remember, the K_a and K_b of a conjugate acid base system is related by the K_w. Using generic terms for acid and base, we adapt the equations above to make one reaction:

$$HA \rightarrow H^+ + A^-$$ plus

$$B + H_2O \rightarrow BH^+ + OH^-$$ will yield:

$$B + HA \rightarrow BH^+ + A^-$$

Water is the solvent in this reaction and is a spectator for this reaction.

Now, when more acid is added to the system, the conjugate base also changes its concentration to maintain the balance provided by the above reaction. So, with small changes, the pH of the solution remains constant. To change the pH by one pH unit, the amount of acid or base added to the system is ten times the initial concentration of the buffer solution. To calculate this, use the Henderson-Hasselbach equation, which is an adaption of the Ka expression.

$$pH = pK_a + \log\left(\frac{[base]}{[acid]}\right)$$

Now there is a buffering capacity present in the solution and when enough acid or base is added, the pH will begin to change dramatically. The higher the concentration of the buffering compound, the stronger the buffer capacity.

Example 6

A 500 mL beaker contains 0.025 moles of acetic acid and 0.020 mole of sodium acetate. Calculate the pH of this solution. For this example, the fact that this solution contains acetic acid and sodium acetate, the conjugate of each other, means that this solution is a buffer. Knowing the names of the anions and the corresponding acid is important for this chapter and a review of chapter 6 in part 1 is recommended.

The concentration of each compound can be calculated, but since the volume is identical for each solute, the moles of each can be used in the equation below.

$$pH = pK_a + \log \frac{base}{acid}$$

$$pH = 4.74 + \log \left(\frac{0.020}{0.025}\right)$$

$$pH = 4.64$$

Example 7

A harder example, 500 mL of water contains 60 g of aniline and 50 g of aniline hydrochloride. Calculate the pH of this solution.

Convert the mass of each solute to moles.

60 g of aniline is 0.644 moles of aniline, which is the base in this example

50 g of aniline hydrochloride is 0.386 moles of aniline hydrochloride, which is the acid in this example

$$pH = pK_a + \log \frac{base}{acid}$$

$$pH = 4.63 + \log \left(\frac{0.644}{0.386}\right)$$

$$pH = 4.85$$

Chapter 18: Oxidation and Reduction

In this book we have been discussing chemical reactions as the transfer of atoms from one molecule to another. However, the electron is matter and the transfer of electrons is considered a chemical reaction. This transfer of electrons describes how many reactions can occur. The following chapter can be confusing, especially the terminology.

What are oxidation and reduction?
To begin, several terms need to be explained. Oxidation is the process of losing electrons. Reduction is the process of gaining electrons. How an element or compound loses electrons is based on the electronegativity of the element taking the electrons and the corresponding electron configurations. Of course, how an element receives extra electrons is also based on its electronegativity compared to the other elements and its electron configurations. Confused? Another way to explain this is that the element that has the highest electronegativity will take the electrons. This is called the oxidizing agent. The element that has the lowest electronegativity will lose the electrons and is called the reducing agent.

To understand which element is undergoing oxidation and reduction, a system of following the electrons is used, oxidation numbers. The rules for determining oxidation numbers are described below:

1. All unbounded elements are given an oxidation number of zero
2. For all unbounded ions, the oxidation number is equivalent to the charge

For the next set of rules, which deal with atoms in a molecule, the assumption is made that bonds are totally ionic, the electrons are owned by the atom with the highest electronegativity.

3. All oxidation numbers in the molecule or ion must equal the charge on the molecule or ion
4. The following ions have a set oxidation number
   a. Hydrogen is 1 or -1.
      i. +1 if bound to a nonmetal
      ii. -1 if bound to a metal
   b. All elements in column 1 are +1
   c. All elements in column 2 are + 2
   d. Boron and aluminum are + 3
   e. Oxygen is -2 or -1
      i. -1 if oxygen is covalently bound to itself (peroxides)

ii. -2 if bound to all other elements except fluoride

iii. If bound to fluoride, oxygen has a positive oxidation number (+2) - this is rare

f. Fluorine is always -1 (highest electronegative element known to man)

g. Elements in column 17 are -1 unless bound to oxygen and fluorine

The maximum range of oxidation numbers, especially valid for this class, is +8 to -8. To determine oxidation numbers in a compound, molecule, or ion, the rules listed above will be used.

Example 1:

Find the oxidation number for each element in NO_3^-

The ion has a total charge of -1, there is one nitrogen and three oxygen atoms. According to the rules in the previous pages, oxygen has an oxidation number of -2. Therefore, $X+(-2*3) = -1$. Solve for X and the oxidation number for nitrogen is 5.

Example 2:

Find the oxidation number for each element in $K_2Fe(CN)_6$

This is a little complicated, so first break apart the ions, which are K and $Fe(CN)_6$. Potassium is always +1 and we have two ions so (2 + anion) = 0. The charge on the anion, $Fe(CN)_6$ is -2. Next step is to break apart Fe from the CN.

The oxidation number on Fe is unknown at this point, but the charge on CN is -1 (you can find this on any list of anions). X - 6 = -2 and the oxidation number for iron is +4. Now carbon and nitrogen. Nitrogen is more electronegative so it will carry the negative oxidation number. The highest negative number nitrogen has is -3, so X - 3 = -1 means that carbon has an oxidation number of +2. Of course, there are several short cuts you can use, but this technique is good for the beginning chemist.

How can electrons move in a reaction?
To determine the movement of electrons in a chemical equation, the student must provide the oxidation number for each element. Then they determine which element is losing electrons (there can be more than one element undergoing oxidation) and which element is gaining electrons, again there can be more than one element undergoing reduction. Now, on paper, divide the equation into two parts, the oxidation reaction and the reducing reaction including the electrons. Then we need to balance each side including the electrons.

How to balance a redox reaction?
The step to balance a redox reaction is very similar to a normal reaction. However, since the number of electrons associated with the elements changes, tracking the movement of the electrons is important. There are many ways to approach these examples,

however, I usually use the following steps to balance these equations.

1. Write the reaction
2. Assign oxidation numbers
3. Determine the oxidation and reduction steps
4. Split the reaction into two
5. Add the electrons needed to reduced or oxidize the element
6. Balance the elements, heaviest first, oxygen and hydrogen last
7. Balance the electrons between the two half reactions
8. Combine the half reactions
9. Add OH- if the solution is basic, leave alone if acidic

The next few examples show how to balance equations of increasing difficulty.

Example 3:

Balance the following redox reaction in acid conditions.

$$Fe_{(s)} + HNO_{3(aq)} \rightarrow Fe^{3+} + NO_3^- + H_{2(g)}$$

Assign oxidation numbers first:

Reactants: Fe = 0, H = +1: NO_3^- does not change in this reaction so the oxidation number for O and N is not relevant for this reaction (for your information, N= 5 and O is -2)

Products: Fe = +3, H=0

Iron is oxidized and hydrogen is reduced. Now split and rewrite the reaction equation into two parts

$$Fe \rightarrow Fe^{3+} + 3\ e^-$$

$$HNO_3 + e^- \rightarrow H_2 + NO_3^-$$

To balance this reaction, three conditions must be met:
1. Masses must be conserved, i.e. the number of atoms in the reactants must be equal to the number of atoms in the product
2. Charge must be conserved, i.e. the total charge in the reactants must be equal to the total charge of the products.
3. Electrons must be conserved, i.e. the number of electrons involved in oxidation must be equal to the number of electrons involved in the reduction.

For the oxidation part listed in the example, mass and charge is conserved. The balancing of electrons will occur shortly. Reviewing the reduction reaction, mass is not conserved. The balanced part of this reaction is written as:

$$2\ HNO_3 + 2\ e^- \rightarrow H_2 + 2\ NO_3^-$$

Now both charge and mass is conserved. However, the electrons involved in the reduction are less than in the oxidation reaction. To make both parts have the same number of electrons, multiply all the components in the oxidation reaction by 2, and multiply all the components in the reduction reaction by 3.

$$2\ Fe \rightarrow 2\ Fe^{3+} + 6\ e^-$$

$$6\ HNO_3 + 6\ e^- \rightarrow 3\ H_2 + 6\ NO_3^-$$

The electrons involved in the oxidation and reduction parts of the reaction are now equal, and the masses and charge are conserved. The last step is to add each part together:

$$2\ Fe_{(s)} + 6\ HNO_{3(aq)} \rightarrow 2\ Fe^{3+} + 6\ NO_3^- + 3\ H_{2(g)}$$

The 6 e- on both sides of the arrow cancel each other out, which is necessary, and now the reaction is balanced.

The next example will show the process when it is necessary to add hydrogen or oxygen to the equation in the form of water.

Example 4:

Balancing redox equation using water.

$$Fe_2O_3 + CO \rightarrow Fe + CO_2$$

Assign oxidation numbers to each element. On the reactant side, carbon is +2, oxygen is -2, and iron is +3. On the product side, carbon is +4, oxygen is -2, and iron is 0. Therefore, carbon is oxidized and iron is reduced.

Split the reaction into the half reactions:

Oxidation: $CO \rightarrow 2\,e^- + CO_2$

Reduction: $Fe_2O_3 + 6\,e^- \rightarrow Fe$

For the oxidation reaction, carbon is balanced and being oxidized from a +2 to +4, so each carbon is losing two electrons. To balance the oxygen, we need to add water to the left side and 2 H+ to the right side of the half reaction.

$$CO + H_2O \rightarrow 2\,e^- + CO_2 + 2\,H^+$$

For the reduction reaction, iron needs to be balanced. Adding another atom of iron to the right side for a total of two will even the iron out. Since iron is being reduced from a +3 to 0, a total of 6 electrons are being transferred in the reduction half reaction. To balance the oxygen, add 3 water molecules to the right side and 6 H+ to the left:

$$Fe_2O_3 + 6\,e^- + 6\,H^+ \rightarrow 2\,Fe + 3\,H_2O$$

To even out the electrons in each half reaction, multiply the oxidation half reaction by 3:

$$3\,CO + 3\,H_2O \rightarrow 6\,e^- + 3\,CO_2 + 6\,H^+$$

Now add the half reactions together to get the final balanced equation:

$$Fe_2O_3 + 3\ CO \rightarrow 2\ Fe + 3\ CO_2$$

Note that water and H⁺ cancel out and are not part of the reaction.

Example 5:

Balance the following reaction occurring in a basic solution.

$$Cr(OH)_3 + Br_{2(l)} \rightarrow CrO_4^{2-} + Br^-$$

First, assign oxidation numbers. Chromium is +3 as a reactant and +6 as a product. Oxygen is always -2 in this reaction and does not change. Bromine is 0 as a reactant and -1 as the product.

Split the reaction into two half reactions. Chromium is being oxidized and bromine is being reduced:

Reduction: $Br_2 \rightarrow Br^-$

Oxidation: $Cr(OH)_3 \rightarrow CrO_4^{2-}$

For the reduction reaction, an additional bromine is added to the products and 2 electrons to the left. The reduction reaction is now balanced. For the oxidation reaction, chromium is balanced. 3 oxygens are on the left and four on the right; 1 water molecule is added to the left and 5 H+ is added to the right, two protons from water and 3 from OH⁻:

Reduction: $Br_2 + 2\ e^- \rightarrow 2\ Br^-$

Oxidation: $Cr(OH)_3 \rightarrow CrO_4^{2-} + 3\ e^-$

The reduction equation is balanced at this point, the oxidation reaction needs an oxygen and a few hydrogens. Add a water molecule to the reagents and 5 H+ to the products:

$$Cr(OH)_3 + H_2O \rightarrow CrO_4^{2-} + 3\ e^- + 5\ H^+$$

At this time, OH- can be added to the equations, however, I prefer to add them last. Next is to balance the electrons by multiplying the reduction reaction by 3 and the oxidation reaction by 2:

Reduction: $3\ Br_2 + 6\ e^- \rightarrow 6\ Br^-$

Oxidation: $2\ Cr(OH)_3 + 2\ H_2O \rightarrow 2\ CrO_4^{2-} + 6\ e^- + 10\ H^+$

Now add the half reactions together:

$$3\ Br_2 + 2\ Cr(OH)_3 + 2\ H_2O \rightarrow 2\ CrO_4^{2-} + 6\ Br^- + 10\ H^+$$

The current equation indicates that the solution is acidic. To make this a basic solution, add 10 OH⁻ to each side of the equation. The following will happen at the products:

$$H^+ + OH^- \rightarrow H_2O$$

So, we have 10 water molecules on the right side and 10 OH- on the left:

$$3\ Br_2 + 2\ Cr(OH)_3 + 2\ H_2O + 10\ OH^- \rightarrow 2\ CrO_4^{2-} + 6\ Br^- + 10\ H_2O$$

Since water is on both sides of the equation, two will cancel leaving 8 water molecules as a product. The final balanced equation is described below:

$$3\ Br_2 + 2\ Cr(OH)_3 + 10\ OH^- \rightarrow 2\ CrO_4^{2-} + 6\ Br^- + 8\ H_2O$$

What is a single displacement reaction?

A single displacement reaction occurs when an element is in contact with a compound that will oxidize the element to form a cation and part of the compound will be reduce to form an element. In this reaction, the cation or a hydronium ion is being replaced with another cation. An example is the oxidation of sodium with water:

$$2\ Na_{(s)} + 2\ H_2O_{(l)} \rightarrow H_{2(g)} + 2\ NaOH_{(aq)}$$

There are several conditions on the atomic level that will drive this reaction. The 2s atomic orbital of sodium has a higher energy level than the 1s orbital of hydrogen and the hydrogen has a slight positive charge due to the electronegativity of the oxygen. So, when sodium is added to water, the hydrogen can pull an electron off sodium and leave the water molecule:

$$Na_{(s)} + H_2O_{(l)} \rightarrow H\cdot + Na^+ + OH^-$$

The reaction above occurs twice to produce two hydrogen radicals. A radical is a compound or an element that has an unpaired electron. Radicals are very reactive. Hydrogen will not be able to react with water molecules or with the hydroxide, but will react with another hydrogen radical to form hydrogen gas:

$$2\,H\cdot \rightarrow H_{2(g)}$$

There are some elements where this reaction is slow. If you review the chapter on kinetics, a technique to speed up the reaction is to increase the concentration of one of the reactants. To achieve this, the metal can be cut up to increase the surface area, or you can add a source of H^+ to the water. Of course, adding an acid will increase the concentration of H^+. The H^+ will pull the valance electron from the metal to form the hydrogen radical and then react with another hydrogen radical to form the hydrogen gas.

There will be a point where adding acid to water will not force the reaction to occur. Again, from the kinetics chapter, another method to increase the kinetics of a reaction is to increase the temperature of the system. Increasing the temperature of the system until the water is near the boiling point will help the hydronium cation react with the metal to dissociate it.

There are some single displacement reactions that can occur in water. These reactions depend on the relative reactivity of the metals to each other. This table describes reactivity of metals. The most reactive is at the top of the list and the reactivity decreases going down the table. The reaction is the oxidation of the metal:

$$metal \rightarrow metal^{n+} + ne^-$$

To use this table, if the cation of the salt is higher than the element listed in the table then the reaction does not proceed. For the single displacement to work, the valance electron of the metal should have higher energy compared to the orbital of the corresponding cation. When this condition is present, the electron can be transferred to the cation and metal is oxidized.

Chapter 19: Electrochemistry and Batteries

There are several different ways to model the transfer of electrons in a reaction. The most common method is to create a galvanic cell. A galvanic cell is one the oldest methods to produce electricity. This technique is a little bulky, think of a car battery. So, in time, scientist and engineers developed a dry cell battery. Then came the alkali battery and finally the lithium/air batteries that we use in most of our electronics. In this chapter, the electrolytic cell will also be discussed, which is a common method to use electricity in the plating of metals on substances.

What is a galvanic cell?

A galvanic cell comprises of two cells connected by a salt bridge and a pair of electrodes. As shown in Figure 19.1, the electrodes are joined together by wires that lead into a voltmeter. In electronics, the battery is connected to the circuit which produces the work. In the example in the figure, one electrode is comprised of copper metal and the other is comprised of iron metal. The cell that contains the copper electrode also contains a solution of copper (II) nitrate. The other cell contains a solution of iron (III) nitrate. The salt bridge is a strip of paper or gel that is saturated with an inert salt, usually potassium nitrate. The purpose of the salt bridge is to facilitate the transfer of charge in the galvanic cell. When the wires are connected and the switch is closed, electrons flow from one electrode through the wires to the other electrode. The electrons flow from high potential to low potential. This potential is the overall energy level that the cells have. At the electrodes, oxidation and reduction occurs. This transfer of electrons will create an imbalance of charge, which is adjusted by the migration of ions in the salt bridge. The electrode where reduction occurs is called the cathode, and the other electrode, where oxidation occurs, is called the anode. The polarity of the electrodes is not an issue for this text, but usually the cathode is negatively charge and the anode is positively charged.

Figure 19.1: Schematic of a galvanic cell

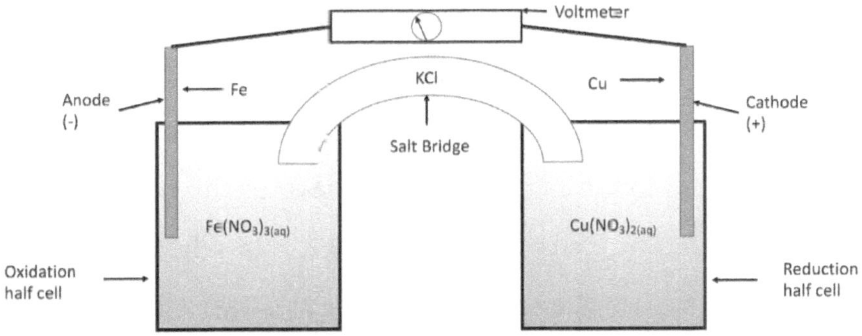

What causes the potential between the electrodes?

The overall energy levels of the atoms in each cell provides the electrical potential of the system. Using the example above, the orbitals of iron cation have a lower electron potential than the copper ions in the other cell. So, iron takes the electrons from the electrode and becomes reduced. The electrons are replaced by the wire which is connected to the other cell. The electrons are taken from the copper electrode and then the copper is dissociated from the electrode into the solution. The electrons from copper are able to leave the copper because the electronic orbitals are higher than the iron's. This difference in potential energies are transferred by the wires to each cell. The change in the potential is instantaneous, but the movement of electrons are not.

How does the salt bridge work?
The salt bridge works by providing ions to balance the charges in the cell. In the left cell, oxidation is taking place and the number of cations increase. To balance the charge, the chloride anions move from the salt bridge into the cell. In the right cell, reduction occurs and an excess negative charge is generated. The potassium ions in the salt bridge move towards the cell to balance the excess positive charge. Figure 19.2 is a schematic of this movement. Now if the salt bridge was not present, the electrons would not move because the potential difference is not continuous. When the salt bridge is in place, the ions in the salt bridge are also under the influence of the electrical potential, and the ions would move to neutralize the charge imbalance. Notice that this movement produces a slight potential between the two

ends of the salt bridge. This imbalance of charge in the salt bridge is called the junction potential and is used to determine the pH of a solution.

Figure 19.2: Schematic of a salt bridge

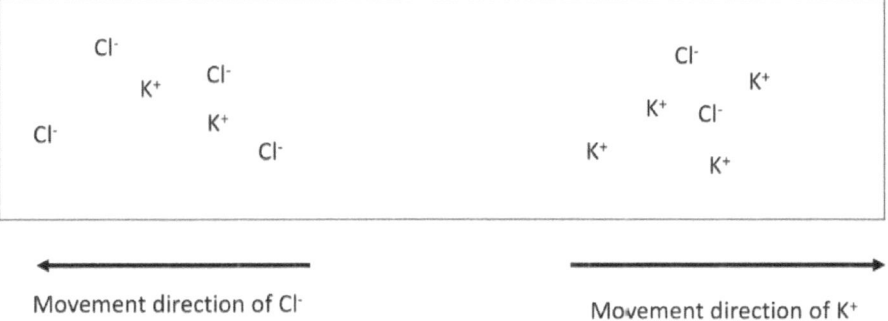

How is electrical potential calculated?

When we place a volt meter between the electrodes, it will measure the difference in the potential. The circuit is not grounded, so either cell can have a certain potential and the difference between the two is measured. Scientists made assumptions to define standard potentials. First, scientists had to determine what conditions needed to be standardized. Concentration of any aqueous solution are a 1 M, gases are at 1 atm or 760 torr, solids and liquids are considered to be at unity, as with equilibrium. The temperature of the system is 298 K to be considered standard conditions. A zero potential was decided and that is the reduction of hydronium ion to hydrogen gas. By definition, this reduction has a potential of zero. Also, the oxidation of hydrogen gas to hydronium is defined to be a zero volts.

$$2 H^+ + 2 e^- \leftrightarrow H_{2(g)}$$

How to write the reactions in the cells.

Instead of drawing a picture each time an electrolytic cell is used, cell notation was developed to write the components of the cell quickly and easily. The line notation starts with the oxidization reactions first. The electrode material is listed first, followed by a solid vertical line. If a solid or gas is in contact with the surface of the electrode, it is listed next. Then the components of the solution are listed and the concentration of the salts are included in parenthesis if known. Next a double vertical line is drawn to indicate a salt bridge. Then the components of the reduction solution are added. Using Figure 19.1, the cell notation for the galvanic cell is:

$$Cu_{(s)} | Cu(NO_3)_{2(aq)} (1.0\ M) || Fe(NO_3)_{3(aq)} (1.0\ M) | Fe_{(s)}$$

As shown in the cell notation above, the concentration of the solutions are listed in the notation. Sometimes the concentrations are omitted depending on the question asked and if the concentrations are known or not. Sometimes the electrode are inert. If this is the case, the electrode will not have a companion ion in solution.

Overall potential difference

If we have all the variables at standard state, and the potential is measured and is positive, the reaction of the cells added together is spontaneous and the cell is galvanic. If the measured voltage is negative, then the reaction is nonspontaneous. Table 19.1 is a list of half reactions and the associated potentials. All ions are in aqueous solutions and the concentrations are at 1 M or the pressure is a 1 atm. The list of reactions are sorted with the elements highly oxidized at the top and the most reduced at the bottom. Notice at the bottom of the table are the elements with a high electronegativity. These elements readily absorb an electron to become an anion and the standard potential reflects this physical property. At the top of the table are electropositive elements that would like to lose electrons. Since the table is the reduction of the element, the potential reflects the physical property and the negative voltage indicates that the reaction would proceed if the equation was written in reverse.

Table 19.1: List of Half Reactions and the Associated Standard Potentials

Reaction	E°
$Li^+ + e^- \leftrightarrow Li$	-3.040 V
$Al(OH)_4^- + 4e^- \leftrightarrow Al + 4 OH^-$	-2.33 V
$V^{2+} + 2e^- \leftrightarrow V$	-1.175 V
$2 H_2O + 2 e^- \leftrightarrow H_2 + 2 OH^-$	-0.8277 V
$Cr^{3+} + 3 e^- \leftrightarrow Cr$	-0.744 V
$Cr^{3+} + e^- \leftrightarrow Cr^{2+}$	-0.407 V
$V^{3+} + e^- \leftrightarrow V^{2+}$	-0.255 V
$2 NO_2^- + 2 H_2O + 4 e^- \leftrightarrow N_2O_2^{2-} + 4 OH^-$	-0.18 V
$Fe^{3+} + 3 e^- \leftrightarrow Fe$	-0.037 V
$2 H^+ + 2 e^- \leftrightarrow H_2$	0.000 V
$AgBr + e^- \leftrightarrow Ag + Br^-$	0.07133 V
$Sn^{4+} + 2 e^- \leftrightarrow Sn^{2+}$	0.151 V
$Cu^{2+} + e^- \leftrightarrow Cu^+$	0.153 V
$Cu^{2+} + 2 e^- \leftrightarrow Cu$	0.3419 V
$Fe^{3+} + e^- \leftrightarrow Fe^{2+}$	0.771 V
$O_2 + 4 H^+ + 4 e^- \leftrightarrow 2 H_2O$	1.229 V
$Au^{3+} + 2 e^- \leftrightarrow Au^+$	1.401 V
$Au^+ + e^- \leftrightarrow Au$	1.692 V
$F_2 + 2 e^- \leftrightarrow 2 F^-$	2.866 V

Table 19.1 is republished with permission of Taylor and Francis Group LLC Books, from CRC Handbook of Chemistry and Physics, 73rd edition, published 1993; permission conveyed through Copyright Clearance Center, Inc. "

What is the relationship between potential and thermodynamics?

In Chapter 13, a spontaneous reaction is a reaction that the change in Gibbs' Free Energy is negative. If the potential of a cell is known, the Gibbs' free energy can be calculated using the equation below:

$$\Delta G = -nFE$$

Where n is the number of electrons being transferred according to the balanced equation, F is Faraday's constant, 96485 C/mol or 96500 C/mol, and E the potential different between the two half cells.

Since we know that Gibbs free energy is also related to the equilibrium constant, we can also relate potential energy to K:

$$\Delta G = RT \ln K$$

Where K is the equilibrium constant, T is the temperature of the cell in Kelvin, and R is the gas constant; 8.314 J/K*mol. The relationship between potential and equilibrium is

$$E = \frac{-RT}{nF} \ln K$$

What happens when the cells are not at standard state?

Up to this point, the galvanic cells are at a standard state, the pressure of gases is 1 atm, the concentration is 1 M, and the temperature is 298 K. However, if the concentrations of the components change, we should see a change in the voltage. So, if E° is standard state, it is also equal to ΔG°. If we adjust the system so it is not in standard state, the ΔG will change using the following equation:

$$\Delta G_f = \Delta G^o - \Delta G$$

Where the subscript f indicates final Gibbs' free energy. Now substitute ΔG = -nFE So we have the following equation:

$$-E_f = -E^o + E$$

Where E_f = final potential, E° is standard potential and E is the change in potential.

Substituting E =(-RT/nF) ln K for the change in potential, we now have the Nernst equation

$$E_{cell} = E^o - \frac{RT}{nF} \ln K$$

Or the expanded form:

$$E_{cell} = \left(E_+^o - \frac{RT}{nF} \ln K\right) - \left(E_-^o - \frac{RT}{nF} \ln K\right)$$

Where E_+ is the cathode potential and E_- indicates the anode potential. K is the equilibrium expression for the half reaction for the appropriate half-cell. In many books, the K is replaced with Q, indicating that the system is not at equilibrium. With all of these equations, there are a couple of facts that you need to know. When E_{cell} = 0, the system is at equilibrium and $E^o{}_{cell}$ = K. Also, if E = 0, then ΔG = 0. The next several pages are example problems for electrochemistry. For these examples, the temperature is assumed to be at 298 K.

Example 1

From the electrochemical cell below, write the half reactions, the net ionic equations, and calculate the E°cell.

$$Pt_{(s)}|Fe^{2+}, Fe^{3+} \, ||Cu^{2+}, Cu^{+}|Pt_{(s)}$$

This example of cell notation is common for galvanic or electrolytic cells. The ions that are involved in oxidation or reduction are listed in solution, not the spectator ions. Also, platinum is inert. The chemist knows this due the absence of platinum ions in solution. Reduction reaction is copper (II) to copper (I) and oxidation is the iron (II) to iron (III), as shown in the half reactions below:

$$oxidation: Fe^{2+} \rightarrow Fe^{3+} + e^{-}$$

$$reduction: Cu^{2+} + e^{-} \rightarrow Cu^{+}$$

Next, ensure each half reaction is balanced, the number of electrons involved in oxidation is equal to the electrons involved in reduction. In this example, everything is balanced. Then we can add the two reactions together to obtain the net ionic equation.

$$Fe^{2+} + Cu^{2+} \longrightarrow Fe^{3+} + Cu^{+}$$

To determine E°cell, look up the E° values for both oxidation and reduction half reaction.

E°₊ = 0.153 V, E°₋ = 0.771 V

Do not change the sign of the values, the reversal of the reduction potential is applied in the equation below:

$E^o_{cell} = E^o_+ - E^o_-$

$E^o_{cell} = 0.153\ V - 0.771\ V = -0.618\ V$

This means that the cell is not spontaneous as written.

Example 2:

Determine the E° from an unbalanced redox reaction

This example shows the method to determine E° of a cell where the electrons in the oxidation and reduction half reaction are not the same. Using the cell describe below, the E°cell will be determined.

$Cu_{(s)}|Cu(NO_3)_{2(aq)}(1.0\ M)||Fe(NO_3)_{3(aq)}(1.0\ M)|Fe_{(s)}$

All reagents and products are at the standard states. The nitrate is a spectator ion for this reaction and will not be considered in the problem.

Oxidation: $Cu_{(s)} \rightarrow Cu^{2+} + 2\ e^-$

Reduction: $Fe^{3+} + 3\ e^- \rightarrow Fe_{(s)}$

Now, since the electrons in each half reaction are different, the half reactions must be multiplied by a factor to obtain the least common multiple. In this case, multiply the oxidation reaction by 3 and the reduction reaction by 2 to obtain the following half reactions:

Oxidation: $3\ Cu_{(s)} \rightarrow 3\ Cu^{2+} + 6\ e^-$

Reduction: $2\ Fe^{3+} + 6\ e^- \rightarrow 2\ Fe_{(s)}$

Combined: $3\ Cu_{(s)} + 2\ Fe^{3+} \rightarrow 2\ Fe_{(s)} + 3\ Cu^{2+}$

$E^o_{cell} = -0.037\ V - 0.3419\ V = -0.3789\ V$

The E° for each half reaction does not change when the half reactions are multiplied. The potential of each cell is based on the overall energy levels of the atomic orbitals, not the actual electrons. Now, when a significant number of products are formed, the equilibrium expression will change, and the change in potential is affected by the electrons transferred in the reaction, as seen in example 3.

Example 3

Using the same cell as example 2, but each concentration is now set to 0.1 M

$Cu_{(s)}|Cu(NO_3)_{2(aq)}(1.0\ M)\|Fe(NO_3)_{3(aq)}(1.0\ M)|Fe_{(s)}$

All reagents and products are at the standard states. The nitrate is a spectator ion for this reaction and will not be considered in the problem.

Oxidation: $Cu_{(s)} \rightarrow Cu^{2+} + 2\ e^-$

Reduction: $Fe^{3+} + 3\ e^- \rightarrow Fe_{(s)}$

Since the concentrations are non-standard, there are two methods a chemist can choose from to calculate the E_{cell}.

Example 3a:

Stating with the Nernst equation:

$$E_{cell} = \left(E_+^o - \frac{RT}{nF}\ln Q_+\right) - \left(E_-^o - \frac{RT}{nF}\ln Q_-\right)$$

Each Q is the equilibrium expression for the half reaction after the electrons have been balanced. Below are the correct half reactions for the example:

Oxidation: $3\ Cu_{(s)} \rightarrow 3\ Cu^{2+} + 6\ e^-$ $Q = [Cu^{2+}]^3$

Reduction: $2\ Fe^{3+} + 6\ e^- \rightarrow 2\ Fe_{(s)}$ $Q = \frac{1}{[Fe^{3+}]^2}$

So n = 6 and substituting the each Q and the constants into the equation, you have the equation below:

$$E_{cell} = \left(E_+^o - \frac{8.314(298)}{6*96500}\ln\left(\frac{1}{[Fe^{3+}]^2}\right)\right) - \left(E_-^o - \frac{8.314(298)}{6*96500}\ln([Cu^{2+}]^3)\right)$$

$$E_{cell} = \left(-0.037 - \frac{8.314(298)}{6*96500}\ln\left(\frac{1}{[0.1]^2}\right)\right) - \left(0.3419 - \frac{8.314(298)}{6*96500}\ln(0.1^3)\right)$$

E_{cell} = - 0.369 V

This method is useful if the cathode or anode is a standard electrode – an electrode that has a specific and unchanging potential over time.

Example 3b

This method is a little easier in some cases. The $E°_{cell}$ can be calculated and the Nernst equation is condensed to the equation below:

$$E_{cell} = \left(E_{cell}^o - \frac{RT}{nF} \ln Q\right)$$

Where Q is the equilibrium expression for the balanced reaction of the cell.

For the example,

$$Q = \frac{[Cu^{2+}]^3}{[Fe^{3+}]^2}$$

Therefore,

$$E_{cell} = \left(E_{cell}^o - \frac{8.314(298)}{6*96500} \ln\left(\frac{[Cu^{2+}]^3}{[Fe^{3+}]^2}\right)\right)$$

and E_{cell} = -0.369 V.

www.ingramcontent.com/pod-product-compliance
Lightning Source LLC
Chambersburg PA
CBHW031435210526
45464CB00005B/2212